动物故事 · 我们身边的动物

波利特的趣味赛跑

猪

[美] 帕姆·斯凯内曼 著
[美] C.A. 诺本斯 绘　　王 姝 译

社
国家一级出版社
全国百佳图书出版单位

吉图字：07-2015-4474

图书在版编目（CIP）数据

我们身边的动物．波利特的趣味赛跑·猪／（美）斯凯内曼著；（美）诺本斯绘；王妹译．— 长春：长春出版社，2016.1
（动物故事）
ISBN 978-7-5445-4173-2

Ⅰ．①我… Ⅱ．①斯… ②诺… ③王… Ⅲ．①猪－儿童读物 Ⅳ．①Q95-49

中国版本图书馆CIP数据核字（2015）第269770号

著　　者 ［美］帕姆·斯凯内曼　　绘　　者 ［美］C.A. 诺本斯
译　　者 王　妹　　责任编辑 张　岚　刘　洋　单紫薇

長春出版社 出版　　吉林省吉广国际广告股份有限公司印制
（长春市建设街1377号）（邮编130061　电话88563443）　　长春出版社发行部发行
787㎜×1092㎜　20开　14.4印张　108千字
2016年1月第1版　2016年1月第1次印刷　　定价：80.00元（全12册）
www.cccbs.net

真实&虚构故事

这套充满奇思妙想的图书专为学龄前后的小读者设计。它以极为巧妙的方式，在引领小读者探索自然科学世界的同时，循序渐进地加强了小读者的阅读能力，并以自如穿梭于真实世界与虚构故事间的形式，启发小读者的思维。

FACT 真实：左侧页面展示了真实照片，以增强小读者的科学认知和对信息文本的理解。

FICTION 虚构故事：右侧页面通过一个有趣的叙事故事，配合奇思妙想的插图，来启发小读者无穷的想象力。

FACT OR FICTION? 真实和虚构故事的页面可以分别进行阅读，通过提问、预测、推断和总结的方式提高理解能力。也可以并排连续阅读，能够帮助小读者发掘和研究不同的写作风格。

真实还是虚构故事？通过有趣的测验，加强小读者对于何为真实何为虚构的理解。

书后特附英文原文和词汇表，并配有外教的纯正发音，以极为便捷的方式让小读者感受地道的英文。

FACT

猪生活在一个大家庭中。雄性的猪被称为公猪，雌性的猪叫作母猪，刚出生的猪叫小猪崽。

波利特·斯万是一头非常快乐的小猪崽。他和他的兄弟姐妹、爸爸妈妈生活在美丽的瓦洛湖畔。

FACT

猪是哺乳动物，它们非常聪明。比猪更聪明的哺乳动物是人类、灵长类动物、鲸和海豚。

波利特自豪地认为他的家庭成员是最聪明的！他的妈妈是位老师，爸爸是飞行员。哥哥霍斯利喜欢数学，弟弟奥克是一个科学奇才，而历史对姐姐苏依来说就像是一个乐园。

FACT

猪也会感到无聊。为了打发时间，它们会玩玩具，保持兴奋状态。

波利特热爱跑步，同时也热爱读书。每天做完作业后，波利特都会和他的小伙伴们在公园里赛跑。他们正在为瓦洛学校趣味赛跑做训练。

FACT

小猪会一直吃到饱了为止。它们的主要食物是谷物、草和植物，它有时也会吃昆虫、蠕虫甚至小动物，如老鼠。

在赛跑比赛前有一个野餐。波利特边啃着西瓜边梦想着得第一。“别吃太多了，”他提醒他的哥哥姐姐们，“这样会让我们的速度变慢！”

FACT

猪的鼻子会帮助它寻找食物。猪用鼻子拱蛆虫来吃。

霍斯利趴在野餐篮旁边吃着水果沙拉。“太好吃了！”他大叫起来。小猪崽们吃了很多。然后他们欣赏趣味赛跑的奖品。那是一支金色的小猪钢笔，用它来写作业一定会拿到 A+ 的成绩。

FACT

猪没有汗腺，所以它们要在泥里翻滚保持凉爽。厚厚的泥浆也会保护它们的皮肤不被昆虫叮咬。

突然，苏依小声说：“我们去泥里滚一下凉快凉快吧！”“好主意！”奥克说。他们急忙跑到湖边，像野猪一样打滚。当他们一路小跑赶回去的时候刚好准备起跑。

FACT

猪跑得很快。它们 1 分钟可以跑 200 多米！

发令员说："预备，跑！"所有的比赛选手都起跑了。波利特跑得比以前要困难许多。妈妈和爸爸为他们的小猪宝宝加油。

FACT

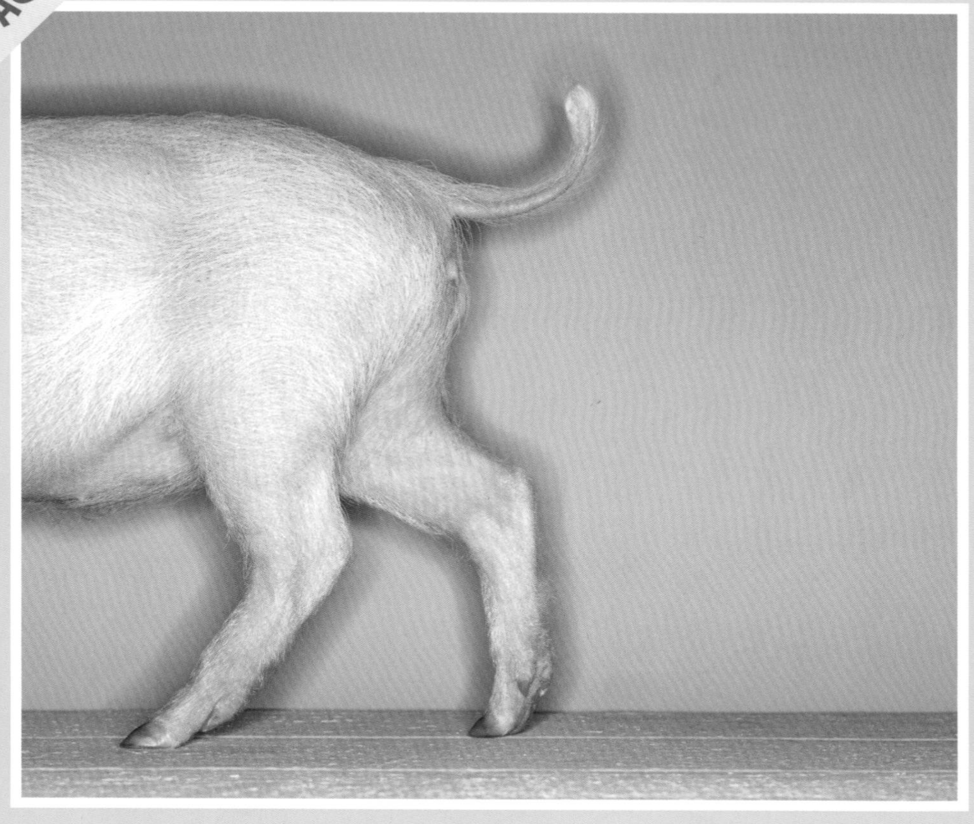

猪走路时只靠每个蹄子上的两根脚趾。它们看起来像是踮着脚尖走路！

FICTION

波利特跑在了第一位，但是当他准备冲线时，霍斯利、苏依和奥克赶上来了。裁判员说："并列第一名，你们都可以获得小猪钢笔！"波利特一边哼着歌一边跳起了舞。他的小伙伴也高兴得跳了起来。

FACT OR FICTION?

阅读下面的句子，判断它们出自真实的部分还是虚构故事的部分。

1. 雌性的猪叫母猪。

2. 猪用钢笔写家庭作业。

3. 猪读历史书。

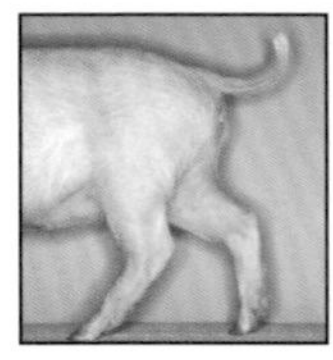

4. 猪走路时只靠每个蹄子上的两根脚趾。

答案

1 真实 2 虚构 3 虚构 4 真实

FACT SPEED READ

Pigs are in the swine family. Male pigs are called 10
boars, female pigs are called sows, and babies are 19
called piglets. 21

Pigs are mammals and are very smart. The only 30
mammals smarter than pigs are humans, primates, 37
whales, and dolphins. 40

Pigs can get bored. To keep from getting bored, 49
pigs play with toys and stay active. 56

Pigs only eat until they are full. Pigs mostly 65
consume grains, grasses, and plants. They will also eat 74
insects, worms, and small animals such as mice. 82

A pig's snout helps it find food. Pigs use their snouts 93
to root for bugs and grubs to eat. 101

Pigs do not have sweat glands, so they roll in the 112
mud to stay cool. The caked-on mud also protects 122
their skin from insect bites. 127

Pigs are fast. They can run a mile in seven minutes! 138

Pigs walk on only two of the four toes on each hoof. 150
It looks like they are walking on tiptoe! 158

FICTION SPEED READ

Riblet Swine is a very happy piglet. He lives with his brothers, sister, mama, and papa in beautiful Wallow Lake Valley.

Riblet proudly thinks that his family is the smartest around! Mama is a teacher and Papa is a pilot. Hogsley loves math, Oink is a science whiz, and history is hog-heaven for Sooey.

Riblet loves running as much as reading. Every day after doing his homework, Riblet races his littermates in the park. They are training for the Wallow School Fun Run.

On the day of the run, there is a picnic before the race. Riblet nibbles watermelon and dreams of winning. “Don’t pig out,” he warns his brothers and sister. “It will slow us down!”

Hogsley roots in the picnic basket for fruit salad. “Yum!” he squeals when he finds it. The piglets eat just enough. Next they admire the

Fun Run prize. It is a golden pig pen, perfect for 154
doing A+ schoolwork. 157

Suddenly Sooey whispers, "Let's roll in the mud 165
to keep cool!" 168

"Great idea!" Oink says. They hurry to the 176
lakeside and wallow like wild boars. They trot back 185
just in time to line up. 191

The starter calls, "Ready, set, go!" The racers take 200
off. Riblet runs harder than he has ever run before. 210
Mama and Papa cheer for their piglets. 217

Riblet takes the lead. But as he nears the finish 227
line, Hogsley, Sooey, and Oink catch up. The referee 236
says, "It's a four-way tie. Golden pig pens for you 247
all!" With a snort, Riblet breaks into a hip-hog 257
dance. His happy littermates join in. 263

GLOSSARY

admire. to regard with pleasure, wonder, and approval

grub. a thick, wormlike insect larva

mammal. a warm-blooded vertebrate that is covered in hair and, in the female, produces milk to feed the young

root. 1) to dig in the dirt with the snout 2) to search for something

snout. the projecting nose or jaws of an animal's head

squeal. to make a high-pitched cry or sound

wallow. to roll around in water, snow, or mud

动物故事 · 我们身边的动物

斯宾塞的网

蜘蛛

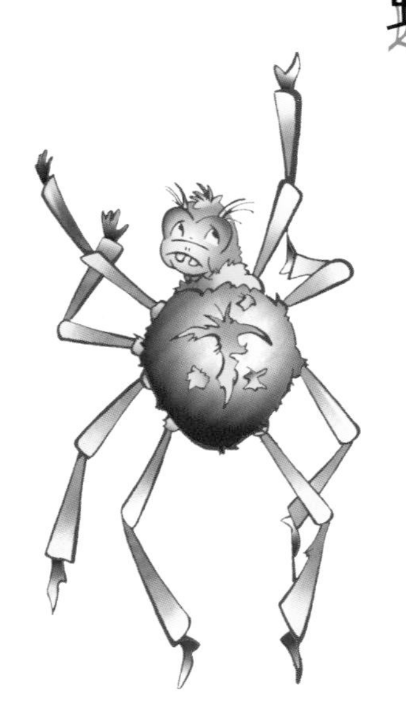

[美] 崔西·康培里恩 著
[美] 安妮·哈伯斯通 绘 王 妹 译

長春出版社
国家一级出版社
全国百佳图书出版单位

吉图字：07-2015-4474

图书在版编目（CIP）数据

我们身边的动物．斯宾塞的网•蜘蛛／（美）康培里恩著；（美）哈伯斯通绘；王妹译．— 长春：长春出版社，2016.1
（动物故事）
ISBN 978-7-5445-4173-2

Ⅰ．①我… Ⅱ．①康… ②哈… ③王… Ⅲ．①蜘蛛目—儿童读物 Ⅳ．①Q95-49

中国版本图书馆CIP数据核字（2015）第270894号

著　　者 ［美］崔西•康培里恩
译　　者 王　妹
绘　　者 ［美］安妮•哈伯斯通
责任编辑 张　岚　刘　洋　单紫薇

長春出版社 出版
（长春市建设街1377号）（邮编130061　电话 88563443）
吉林省吉广国际广告股份有限公司印制
长春出版社发行部发行
787㎜×1092㎜　20开　14.4印张　108千字
2016年1月第1版　2016年1月第1次印刷
定价：80.00元
（全12册）
www.cccbs.net

真实&虚构故事

这套充满奇思妙想的图书专为学龄前后的小读者设计。它以极为巧妙的方式，在引领小读者探索自然科学世界的同时，循序渐进地加强了小读者的阅读能力，并以自如穿梭于真实世界与虚构故事间的形式，启发小读者的思维。

FACT 真实：左侧页面展示了真实照片，以增强小读者的科学认知和对信息文本的理解。

FICTION 虚构故事：右侧页面通过一个有趣的叙事故事，配合奇思妙想的插图，来启发小读者无穷的想象力。

FACT OR FICTION? 真实和虚构故事的页面可以分别进行阅读，通过提问、预测、推断和总结的方式提高理解能力。也可以并排连续阅读，能够帮助小读者发掘和研究不同的写作风格。

真实还是虚构故事？通过有趣的测验，加强小读者对于何为真实何为虚构的理解。

书后特附英文原文和词汇表，并配有外教的纯正发音，以极为便捷的方式让小读者感受地道的英文。

FACT

蜘蛛腿上的毛发和爪子能让它们紧贴蜘蛛网。它们身上的油也能帮助它们不被粘在网上。

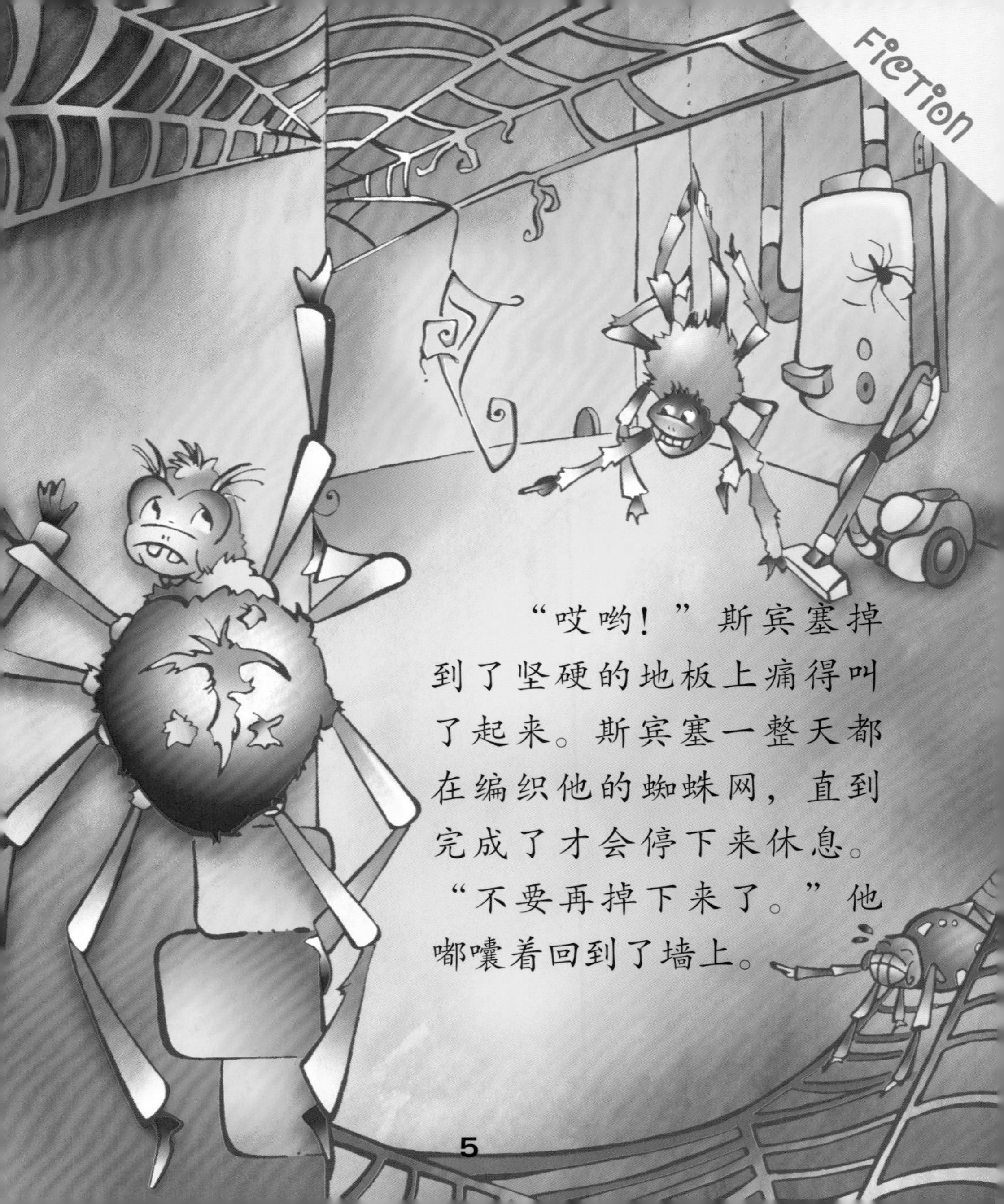

“哎哟！”斯宾塞掉到了坚硬的地板上痛得叫了起来。斯宾塞一整天都在编织他的蜘蛛网，直到完成了才会停下来休息。“不要再掉下来了。”他嘟囔着回到了墙上。

FACT

蜘蛛用几种不同的蜘蛛丝来制作它们的网。它们先搭起一个框架，然后再填充里面。

斯宾塞喜欢编织美丽的蜘蛛网。他是地下室里最具有创造力的蜘蛛。但不幸的是，他的蜘蛛网不久就会被清理掉。

FACT

蜘蛛会回收它们的网。它们吃掉旧的蜘蛛网，以便产生更多的丝来编织新网。

“斯宾塞这么努力地在编织他的网。难道他不知道星期二是清洁日吗？”其他蜘蛛说。

FACT

蜘蛛网上的振动可以让蜘蛛知道是否已经成功捕获猎物。

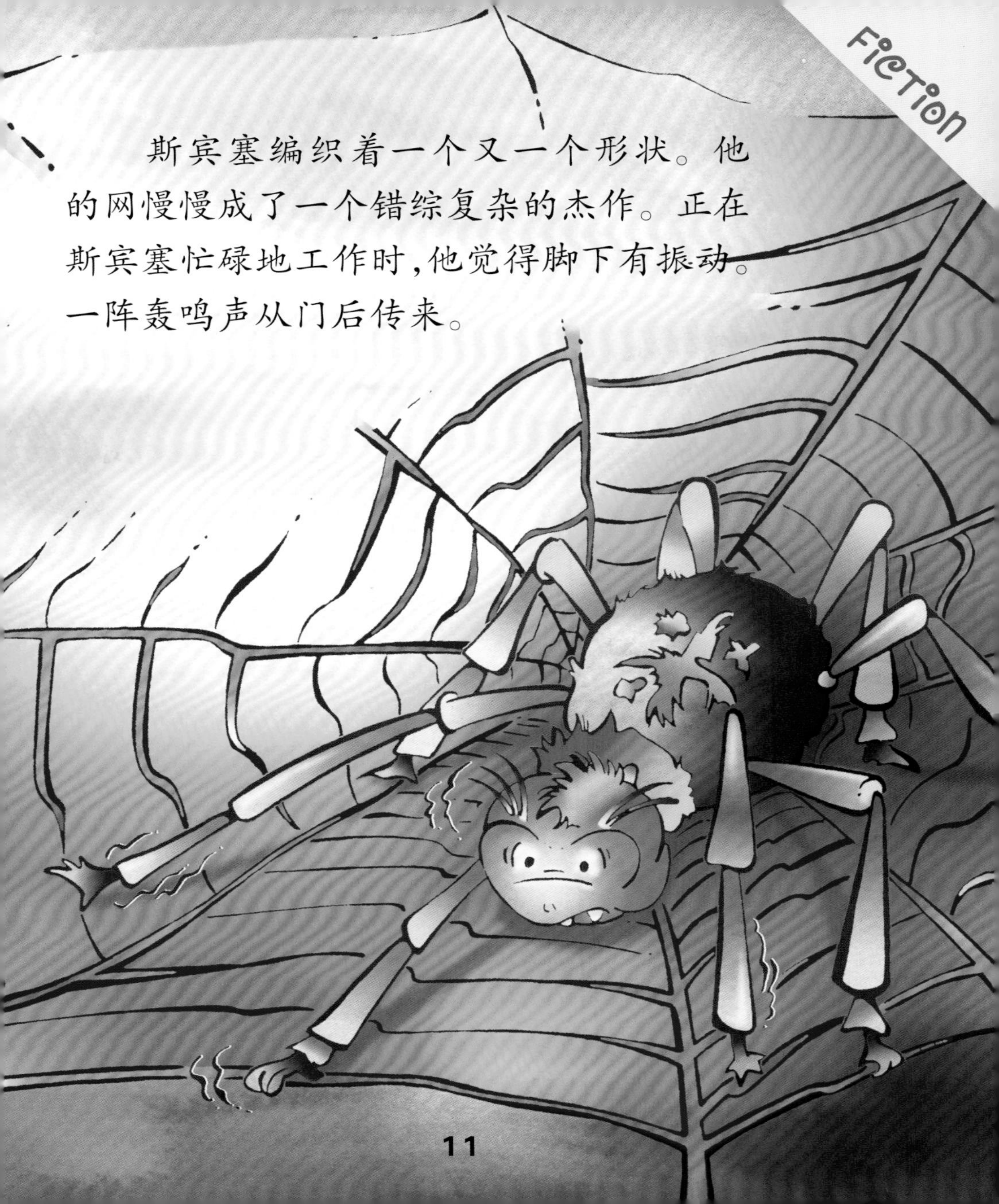

斯宾塞编织着一个又一个形状。他的网慢慢成了一个错综复杂的杰作。正在斯宾塞忙碌地工作时,他觉得脚下有振动。一阵轰鸣声从门后传来。

FACT

蜘蛛会用丝做成一张网。一只蜘蛛大约需要花三个小时来完成一张蜘蛛网。

随着门的打开，斯宾塞从空中掉到了地板上。他抬起头正好看到一个塑料管朝他的蜘蛛网挥去。“不要，”他哭着喊，“我完美的蜘蛛网！”斯宾塞的网在一瞬间就被吸进管子里了。

FACT

蜘蛛能适应许多环境下的生活。它们可以在地上、在树上、在植物上、在洞穴里，甚至在水上生存。

斯宾塞开始觉得他永远都做不出一张能够长存的蜘蛛网了。“我必须比以前加入更多的创意。”斯宾塞暗下决心。在他下一次的尝试中，他发现了一个新的地方。其他蜘蛛对斯宾塞的决心表示深深的尊敬。

FACT

蜘蛛丝很有弹性也很强韧。一些蜘蛛丝甚至跟一根直径相同的钢铁具有同样的强度。

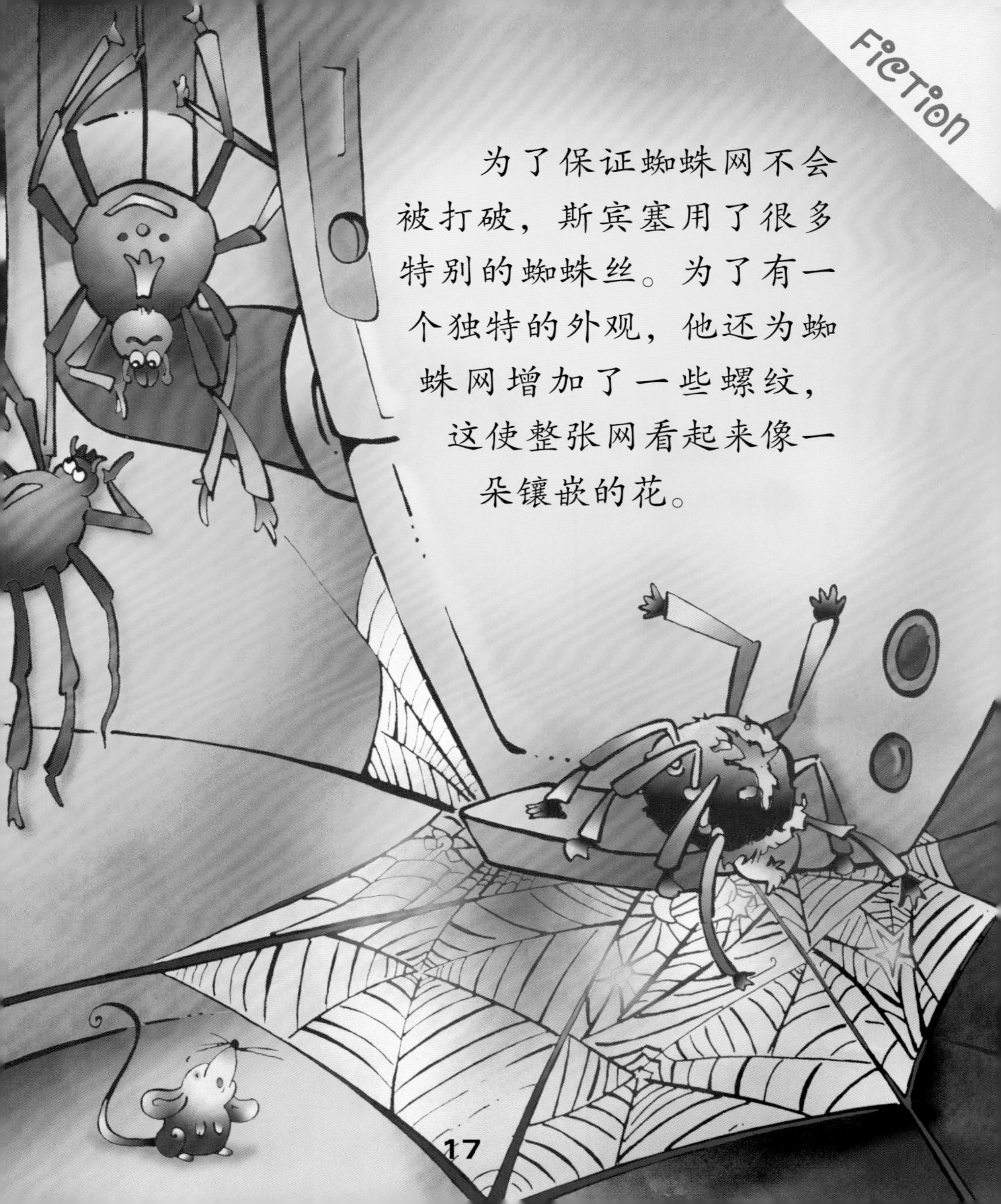

为了保证蜘蛛网不会被打破，斯宾塞用了很多特别的蜘蛛丝。为了有一个独特的外观，他还为蜘蛛网增加了一些螺纹，这使整张网看起来像一朵镶嵌的花。

FACT

世界上大约有 35000 种蜘蛛。

蜘蛛网完成后，斯宾塞邀请所有住在地下室的蜘蛛来聚会。起初他们有些担心，但是这张网完全能够容纳整个派对！蜘蛛们都祝贺斯宾塞建造了一张美丽又坚固的网。

FACT OR FICTION?

阅读下面的句子，判断它们出自真实的部分还是虚构故事的部分。

1. 蜘蛛可以说话。

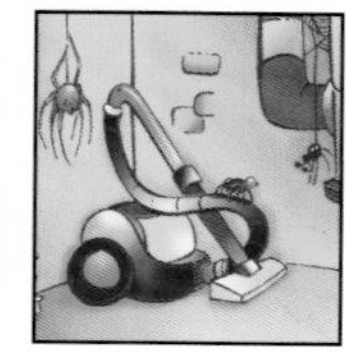

2. 蜘蛛能在许多环境下生活。

3. 蜘蛛丝很强韧。

4. 蜘蛛会举行聚会。

答案

1 虚构 2 真实 3 真实 4 虚构

FACT SPEED READ

The hair and claws on spiders' legs allow them to cling to their webs. The oils on their bodies keep them from sticking to their own webs.

Spiders make their webs from several different kinds of spider silk. They make the framework first, then fill it in.

Spiders are recyclers. They eat their old webs to make more silk for new webs.

Vibrations on the web let a spider know it has caught an insect.

A spider uses from 22 to 66 yards of silk to make a web. It takes a spider about three hours to spin a web.

Over time, spiders have adapted to live in many habitats. They can be found on the ground, in trees and plants, in caves, and even on the water.

Spider silk is very elastic and strong. Some spider silks are stronger than a strand of steel that is the same diameter.

There are about 35,000 known species of spiders.

10 21 27 36 47 56 62 73 75 88 100 109 119 128 137 149 150 158

FICTION SPEED READ

"Ouch!" Spencer screams as he hits the hard floor. Spencer worked on his web all day, only for it to break as he was finishing. "Not again," he mumbles as he makes his way back up the wall.

Spencer loves to make beautiful webs. He is the most creative spider in the basement. Unfortunately, he has trouble making a web that lasts.

"Spencer is working so hard on his web. Doesn't he know Tuesday is cleaning day?" the other spiders say.

Spencer creates shape after shape. His web is becoming an intricate masterpiece. As Spencer works, he feels a vibration under his feet. A loud roar comes from behind the door.

As the door swings open, Spencer tumbles through the air and lands on the floor. He looks up just in time to see a plastic tube heading for

his web. "No," he cries. "Not my perfect web!" 148
Spencer's web is sucked up in an instant. 156

Spencer is starting to think that he will never 165
make a durable web. "I will have to be more 175
creative than ever before," Spencer decides. For his 183
next attempt, he finds a new spot under the 192
furnace. The other spiders are in awe of Spencer's 201
determination. 202

To make sure this web won't break, Spencer takes 211
the time to use extra silk. For a unique look, he 222
adds threads that make the web look like a mosaic. 232

Once the web is finished, Spencer invites all of the 242
basement spiders over. They are hesitant, but the web 251
holds the entire party! The spiders all congratulate 259
Spencer on his beautiful and sturdy web. 266

GLOSSARY

creative. having the ability to make something original

determined. having made a firm decision about something

durable. long lasting and able to withstand wear

habitat. the area or environment where a person or thing usually lives

intricate. having parts that are arranged in a complex or elaborate manner

mosaic. a decorative design made up of many small parts

unique. the only one of its kind

vibration. the act of moving rapidly back and forth

英雄格兰特

蚂蚁

[美] 安德斯·汉森 著
[美] 安妮·哈伯斯通 绘　王妹 译

長春出版社
国家一级出版社
全国百佳图书出版单位

吉图字：07-2015-4474

图书在版编目（CIP）数据

我们身边的动物．英雄格兰特·蚂蚁／（美）汉森著；（美）哈伯斯通绘；王妹译．—长春：长春出版社，2016.1
（动物故事）
ISBN 978-7-5445-4173-2

Ⅰ．①我… Ⅱ．①汉… ②哈… ③王… Ⅲ．①蚁科—儿童读物 Ⅳ．①Q95-49

中国版本图书馆CIP数据核字（2015）第270893号

著　　者 [美]安德斯·汉森　　绘　　者 [美]安妮·哈伯斯通
译　　者 王　妹　　责任编辑 张　岚　刘　洋　单紫薇

長春出版社 出版　　吉林省吉广国际广告股份有限公司印制
（长春市建设街1377号）（邮编130061　电话 88563443）　　长春出版社发行部发行
787㎜×1092㎜　20开　14.4印张　108千字
2016年1月第1版　2016年1月第1次印刷
定价：80.00元
（全12册）

www.cccbs.net

真实&虚构故事

这套充满奇思妙想的图书专为学龄前后的小读者设计。它以极为巧妙的方式，在引领小读者探索自然科学世界的同时，循序渐进地加强了小读者的阅读能力，并以自如穿梭于真实世界与虚构故事间的形式，启发小读者的思维。

FACT 真实：左侧页面展示了真实照片，以增强小读者的科学认知和对信息文本的理解。

FICTION 虚构故事：右侧页面通过一个有趣的叙事故事，配合奇思妙想的插图，来启发小读者无穷的想象力。

FACT OR FICTION? 真实和虚构故事的页面可以分别进行阅读，通过提问、预测、推断和总结的方式提高理解能力。也可以并排连续阅读，能够帮助小读者发掘和研究不同的写作风格。

真实还是虚构故事？通过有趣的测验，加强小读者对于何为真实何为虚构的理解。

书后特附英文原文和词汇表，并配有外教的纯正发音，以极为便捷的方式让小读者感受地道的英文。

FACT

工蚁是没有翅膀的蚂蚁，它们也不能繁衍后代。

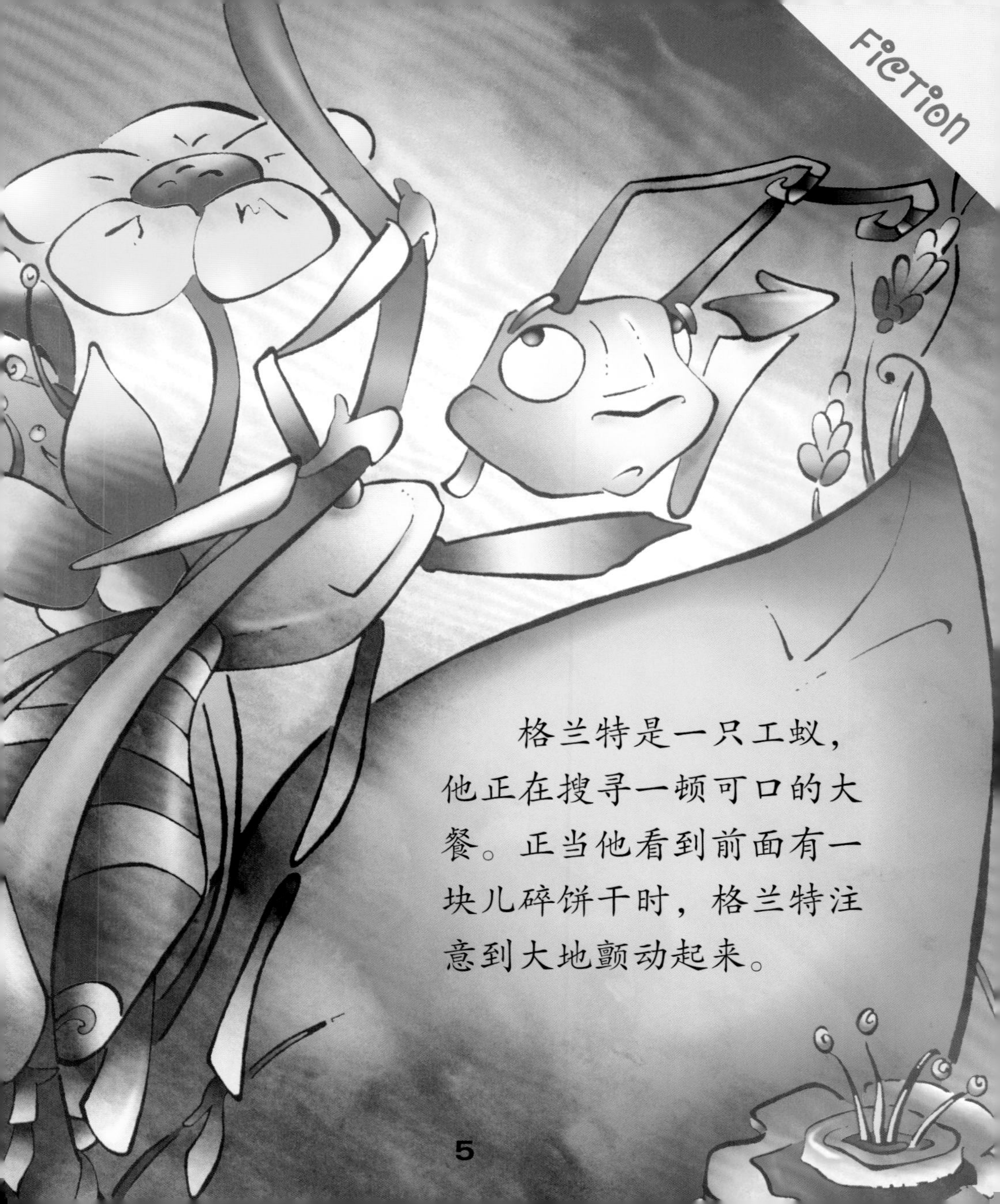

格兰特是一只工蚁，他正在搜寻一顿可口的大餐。正当他看到前面有一块儿碎饼干时，格兰特注意到大地颤动起来。

FACT

蚂蚁可以用长在它们腿上、触角上、胸部和头部的器官来检测地面上的震动。

格兰特年轻的时候曾经听过一个可怕的故事，故事讲了一个巨大的金属怪物撼动大地。他有一种奇怪的感觉，仿佛就是那个巨大的怪物引起了地面的震动。

FACT

由许多只蚂蚁的家堆砌成的土丘，被叫作蚁丘。

格兰特赶紧回家报告女王："尊敬的女王大人，我有一个坏消息要告诉您。恐怕我们就要遭到最大的敌人——那个无坚不摧的推土机布尔的攻击了！"

FACT

蚂蚁会通过分泌化学物质引导其他蚂蚁做出具体的行动，或者通过互相接触来进行沟通。

“这不可能！”女王惊讶地说道。“千万不要是推土机布尔！拉响警报！警告我的女仆们！我们要立即离开这里！”

大家纷纷行动起来。几乎每个人都跳了起来，向四面八方涌动！

FACT

有一些种类的蚂蚁会繁衍出较大的工蚁，这些工蚁是蚂蚁中的士兵，它们的工作是保护蚁巢。

正在蚂蚁们向出口逃跑时，格兰特听到女王向混乱中的蚂蚁们大声命令：“保持秩序和队列！士兵在前方开路，工兵紧随其后。准备出发！”

FACT

如果大量蚂蚁被一次性杀死，那么附近的蚂蚁就会闻到死蚂蚁所释放的化学物质，并发动疯狂的进攻。

但是正当他们要离开时，推土机布尔巨大的黄色头部越过灌木丛，开始驶向格兰特！

FACT

雄性蚂蚁的主要作用是繁殖后代，在这之后它们很快就会死去。

格兰特正准备要逃跑，他看到周围的人指着天空欢呼。“是飞行员！他们来救我们了！”飞行队的飞蚁们盘旋在空中，每人手里都拿着长长的绳子。

FACT

如果没有蚁后，一个蚁群不能繁殖更多的蚂蚁，因此就不会长久存在下去。

一只飞蚁飞过格兰特的上空，格兰特紧紧抓住了绳子。飞蚁拉起格兰特，和其他同胞一起飞向高空，把所有蚂蚁带到了很远的地方，女王正在那里等待他们。

“谢谢你，格兰特。”她说，“你的警报救了我的子民。为此，我授予你蚂蚁之王的荣誉！”

FACT OR FICTION?

阅读下面的句子，判断它们出自真实的部分还是虚构故事的部分。

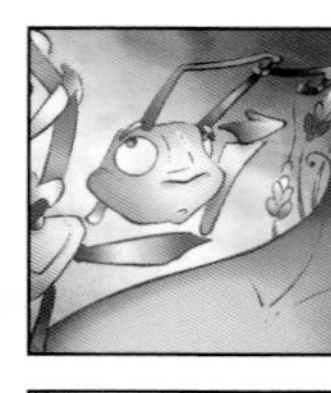

1. 工蚁无法繁殖后代。

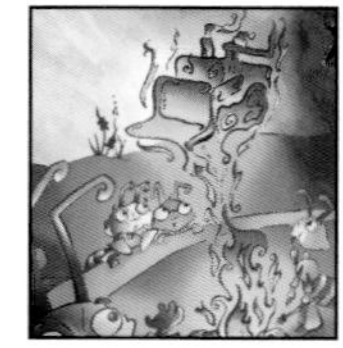

2. 年轻的蚂蚁听说过关于推土机的可怕故事。

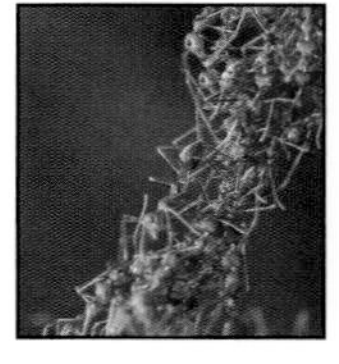

3. 蚂蚁通过释放化学物质来交流。

4. 飞蚁在飞行时带着绳子。

答案

1 真实 2 虚构 3 真实 4 虚构

FACT SPEED READ

Worker ants are wingless ants that cannot reproduce. 8

Ants can detect vibrations in the ground with organs on their legs, antennae, thorax, and head. 16 24

The entries to many ant homes are marked by mounds called anthills. 33 36

Ants can communicate by secreting chemicals that trigger specific behaviors in other ants or by touching each other. 43 52 54

Some ant species produce larger worker ants, called soldiers, whose job is to protect the nest. 62 70

If many ants are killed at once, nearby ants may smell chemicals released by the dead ants and launch a frenzied attack. 80 89 92

The primary purpose of male ants is reproduction, after which they soon die. 100 105

Without a queen ant, an ant colony can't produce more ants and therefore won't last long. 114 121

FICTION SPEED READ

Grant, a worker ant, is in search of a treat. Just as Grant spies a crumbled cookie up ahead, he notices the earth is trembling.

As a young ant, Grant heard terrible tales of a massive metal monster that shook the earth. He has an eerie feeling that it's this huge monster that's causing the ground to shake.

Grant hurries back home to address the queen, "My Lady, I have grim news. I fear we are under attack by none other than our archenemy, the indomitable Bull Dozer!"

"It can't be!" the queen gasps. "Not Bull Dozer! Sound the alarm! Alert my handmaidens! We leave at once!"

There is a flurry of activity. Everybody seems to jump up and jet in every direction at once!

As a river of ants streams out the exits, Grant can hear the queen barking orders amid the

confusion, "Rank and file! Soldiers to the front, 151
workers fall back. Prepare to march!" 157

But just as they are leaving, Bull Dozer's big 166
yellow head bursts through the bushes and begins 174
to barrel right toward Grant! 179

As Grant turns to run, he notices others around 188
him pointing to the sky and cheering. "It's the 197
flyers! They've come to save us!" one of them 206
exclaims. Every winged ant in the colony is in flight 216
and holding a long piece of string. 223

As a flyer passes over Grant, he grabs the string 233
and holds tight. The flyers lift Grant and his 242
comrades high into the air and carry them to a 252
distant field where the Queen awaits them. 259

"Thank you, Grant," she says. "Your warning has 267
saved our colony. For this, I dub thee Grant, King of 278
the Ants!" 280

GLOSSARY

comrade. a friend or a fellow member of a group

dub. to give a special title or name to someone

frenzied. showing uncontrolled excitement or agitation

handmaiden. a female servant

indomitable. impossible to conquer

secrete. to form and give off a specific substance

species. a group of organisms with similar characteristics

thee. another word for *you* most often used in religious or literary works

thorax. the section of a body that is between the head and the abdomen

雷克斯的美容沙龙

鸡

[美] 凯利·杜德娜 著
[美] 安妮·哈伯斯通 绘 王 妹 译

長春出版社
国家一级出版社
全国百佳图书出版单位

吉图字：07-2015-4474

图书在版编目（CIP）数据

我们身边的动物．雷克斯的美容沙龙·鸡 /（美）杜德娜著；（美）哈伯斯通绘；王妹译．— 长春：长春出版社，2016.1
（动物故事）
ISBN 978-7-5445-4173-2

Ⅰ．①我… Ⅱ．①杜… ②哈… ③王… Ⅲ．①鸡—儿童读物 Ⅳ．①Q95-49

中国版本图书馆 CIP 数据核字（2015）第 269703 号

著　　者 [美] 凯利·杜德娜　　绘　　者 [美] 安妮·哈伯斯通
译　　者 王　妹　　责任编辑 张　岚　刘　洋　单紫薇

長春出版社 出版　　吉林省吉广国际广告股份有限公司印制
（长春市建设街 1377 号）（邮编 130061　电话 88563443）　　长春出版社发行部发行
787 mm×1092 mm　20 开　14.4 印张　108 千字
2016 年 1 月第 1 版　2016 年 1 月第 1 次印刷
定价：80.00 元
（全 12 册）

www.cccbs.net

真实&虚构故事

这套充满奇思妙想的图书专为学龄前后的小读者设计。它以极为巧妙的方式，在引领小读者探索自然科学世界的同时，循序渐进地加强了小读者的阅读能力，并以自如穿梭于真实世界与虚构故事间的形式，启发小读者的思维。

FACT 真实：左侧页面展示了真实照片，以增强小读者的科学认知和对信息文本的理解。

FICTION 虚构故事：右侧页面通过一个有趣的叙事故事，配合奇思妙想的插图，来启发小读者无穷的想象力。

FACT OR FICTION? 真实和虚构故事的页面可以分别进行阅读，通过提问、预测、推断和总结的方式提高理解能力。也可以并排连续阅读，能够帮助小读者发掘和研究不同的写作风格。

真实还是虚构故事？通过有趣的测验，加强小读者对于何为真实何为虚构的理解。

书后特附英文原文和词汇表，并配有外教的纯正发音，以极为便捷的方式让小读者感受地道的英文。

FACT

公鸡不只在黎明时分打鸣，它们在一天中的任何时间都会打鸣。鸡鸣是公鸡确定自己领土的一种方式。

早上 6 点，公鸡雷克斯的闹钟响了。他按下“打盹儿”按钮，心想：“只要再睡 10 分钟，然后我就起床打鸣。”

FACT

如果一群鸡中公鸡不在，母鸡就会停止产蛋，开始鸣叫。

6点10分，

雷克斯的闹钟又响了。

他对他的妻子母鸡海伦说：

“亲爱的，你替我去完成打鸣这个任务，好吗？我并不是懒，只是我不是一只晨鸡。”

FACT

鸡在白天活动。它们晚上不活动，一起在栖木上休息。

雷克斯想换个不同的工作，他决定开一家美容沙龙。“我自己就是老板了。”他高兴地说，“要是我不愿意早起，那就可以中午再去上班！”

FACT

母鸡和公鸡的头顶上都有鸡冠。但是公鸡的鸡冠会更大。

雷克斯擅长给鸡冠做造型。他在墙上挂了一张海报，这样他的客户就可以看到并能选择不同的鸡冠风格了。

FACT

鸡通过相互啄咬来分辨它们的等级。这就是啄序（社会等级）的来源。处在更高阶级的鸡会有权获得更好的食物和筑巢地点。

店铺开业的第一天，雷克斯带着他的家人和朋友首先进行体验。外公拉尔夫坐在第一把椅子上说：“雷克斯，我不是年轻人，不需要漂亮的中分，给我做一个简单的造型就可以。”

FACT

母鸡的呼吸频率几乎比公鸡快一倍。

雷克斯的下一个客户是海伦。她兴奋地急促呼吸着。她问道：“雷克斯，你觉得金凤花式的造型会好看吗？”

雷克斯咯咯地说：“亲爱的，你什么造型都很好看！”

FACT

鸡冠是鸡展示自身容貌的重要身体部分，并且能帮助鸡保持凉爽。

海伦把他们的儿子吉普也带来了，这是他第一次做中分造型。雷克斯跟吉普说：“儿子，我觉得不管是核桃还是草莓风格都很酷。你喜欢哪一种？”

“哦，爸爸！请给我做一个核桃风格的吧！”

FACT

世界上大约有 175 种不同的鸡，分属 60 个不同的类别。

在第一周结束时，雷克斯已经为60只鸡做了造型。他锁上门，对自己说：“太累了！也许天亮时打鸣并不是最糟的工作！”

FACT OR FICTION?

阅读下面的句子，判断它们出自真实的部分还是虚构故事的部分。

1. 打鸣是公鸡确定领土的一种方式。

2. 鸡用闹钟来叫醒自己。

3. 鸡会去美容沙龙做鸡冠造型。

4. 鸡相互啄咬来分辨等级。

答案

1 真实 2 虚构 3 虚构 4 真实

FACT SPEED READ

Roosters crow not only at dawn but at any time of 11
day. Crowing is one way roosters define their territory. 20

If a rooster is not present in a flock of chickens, a 32
hen might stop laying eggs and begin to crow. 41

Chickens are daytime animals. At night they are 49
inactive and roost close together on a perch. 57

Both hens and roosters have combs on the tops of 67
their heads. However, roosters' combs are larger. 74

Chickens peck each other to figure out their rank. 83
That's where the phrase pecking order comes from. 91
Chickens higher in the order have better access to 100
food and nesting locations. 104

Hens breathe almost twice as fast as roosters. 112

Chickens' combs are a showy body part, but they 121
also help chickens stay cool. 126

There are about 175 varieties of chickens, which are 135
grouped into about 60 breeds. 140

FICTION SPEED READ

It's 6:00 a.m., and Rex Rooster's alarm clock goes off. He hits the snooze button and thinks to himself, "Just 10 more minutes. Then I'll get up and crow."

It's 6:10 a.m., and Rex's alarm clock goes off again. He says to his wife, Helen Hen, "Honey, would you take over the crowing duties, please? I'm not lazy, but I'm just not a morning chicken."

Rex thinks about different jobs and decides to open a beauty salon. "I'm my own boss now," he says happily. "I don't have to go to work until noon if I don't want to!"

Rex specializes in comb styling. He hangs a poster on the wall so that his customers can see what styles are available.

During his first day of business, Rex takes his family and friends ahead of everyone else. Grandpa Ralph Rooster sits in the first chair and

says, “Rex, I’m no spring chicken, and I don’t need 156
a fancy comb-over. Just give me the single.” 165

Rex’s next customer is Helen. She is breathless 173
with excitement. She says, “Rex, do you think the 182
buttercup would look good on me?” 188

Rex clucks, “Honey, everything looks good on 195
you!” 196

Helen has also brought their son Chip Chick for 205
his first comb-out. Rex says to Chip, “Son, I think 216
either the walnut or the strawberry would be a cool 226
style for you. Which one would you like?” 234

“Oh, Dad! Give me the walnut, please!” Chip 242
exclaims. 243

By the end of the first week, Rex has styled the 254
combs of 60 chickens. He locks the door and says to 265
himself, “I’m exhausted! Maybe crowing at dawn 272
wasn’t such a bad job after all!” 279

GLOSSARY

access. the right to use something, enter a place, or talk to someone

comb. the fleshy crest on the head of a bird

crow. to make the loud, shrill cry of a rooster

flock. a group of animals or birds that have gathered or been herded together

hen. an adult female chicken

rank. one's social class or position in a group

roost. to sit or sleep on a perch

rooster. an adult male chicken

spring chicken. 1) a young chicken 2) a slang expression meaning *young person*

动物故事 · 我们身边的动物

玛丽的账单

鸭

［美］ 崔西·康培里恩 著

［美］ 安妮·哈伯斯通 绘 王 妹 译

長春出版社

国家一级出版社

全国百佳图书出版单位

吉图字：07-2015-4474

图书在版编目（CIP）数据

我们身边的动物．玛丽的账单·鸭 /（美）康培里恩著；（美）哈伯斯通绘；王妹译．— 长春：长春出版社，2016.1

（动物故事）

ISBN 978-7-5445-4173-2

Ⅰ．①我…　Ⅱ．①康…　②哈…　③王…　Ⅲ．①鸭科－儿童读物　Ⅳ．①Q95-49

中国版本图书馆 CIP 数据核字（2015）第 269746 号

著　　者 ［美］崔西·康培里恩　　绘　　者 ［美］安妮·哈伯斯通

译　　者 王　妹　　责任编辑 张　岚　刘　洋　单紫薇

長春出版社 出版　　吉林省吉广国际广告股份有限公司印制

（长春市建设街 1377 号）（邮编 130061　电话 88563443）　　长春出版社发行部发行

787 mm×1092 mm　20 开　14.4 印张　108 千字

2016 年 1 月第 1 版　2016 年 1 月第 1 次印刷

定价：80.00 元（全 12 册）

www.cccbs.net

真实&虚构故事

这套充满奇思妙想的图书专为学龄前后的小读者设计。它以极为巧妙的方式，在引领小读者探索自然科学世界的同时，循序渐进地加强了小读者的阅读能力，并以自如穿梭于真实世界与虚构故事间的形式，启发小读者的思维。

FACT 真实：左侧页面展示了真实照片，以增强小读者的科学认知和对信息文本的理解。

FICTION 虚构故事：右侧页面通过一个有趣的叙事故事，配合奇思妙想的插图，来启发小读者无穷的想象力。

FACT OR FICTION? 真实和虚构故事的页面可以分别进行阅读，通过提问、预测、推断和总结的方式提高理解能力。也可以并排连续阅读，能够帮助小读者发掘和研究不同的写作风格。

真实还是虚构故事？通过有趣的测验，加强小读者对于何为真实何为虚构的理解。

书后特附英文原文和词汇表，并配有外教的纯正发音，以极为便捷的方式让小读者感受地道的英文。

FACT

破壳而出的小野鸭们那柔软的绒毛一变干，鸭妈妈就将它们带下水了。

野鸭玛丽正在湖里游泳，想着自己的事儿，这时她注意到了天空中有一个巨大的灰色物体。

这些年幼的，只有绒毛的小野鸭们还不能飞，要等它们长到两个月左右，羽毛丰满的时候才行。

“噢，我的天，那是什么？”她惊呼道。

“是飞机。人类可以乘它去很远的地方。”鸭爸爸德雷克说，“你很快就能学会像它那样飞啦。”

FACT

野鸭的骨头是中空的，这可以让它们的身体更轻，让远距离飞行更容易些。

玛丽很惊讶！她现在就想学飞！玛丽的爸爸给她做示范，教她如何飞。玛丽也很用心地练习，直到她能飞得很高很高。

FACT

在秋天的时候，野鸭们会为它们的身体储备充足的脂肪，为它们飞往南方过冬的旅程提供能量。

在这一天结束的时候，玛丽拍打翅膀的飞行练习使她筋疲力尽。和这么辛苦的练习相比，她觉得搭乘飞机听起来是个更棒的主意！

FACT

野鸭以湿地的植物和谷物为食，如小麦、大麦和燕麦。它们也吃一些昆虫和贝类水生动物。

第二天，玛丽前往机场，登上了飞机。玛丽刚在座位上坐好，乘务员就过来问她："今天您的午饭想要卤汁面条还是牛排？"玛丽两样都点了！

FACT

野鸭在北美和加拿大度过春季和夏季，它们在冬天飞回南方。

几小时后，飞机着陆了，玛丽兴奋地畅想在德克萨斯州的温暖天气中尽情享受。在下飞机的时候，乘务员递给玛丽一个信封，信封里是一张 200 美元的账单！

FACT

鸭子在走路的时候总是摇摇摆摆，这是因为它们的双腿短小且分得很开。

Fiction

玛丽用她全部的钱付了账单，她非常沮丧地走下飞机时，已经身无分文了。

FACT

野鸭们不常独自飞行，它们总是成群结队地飞。一队鸟儿组成一个鸟群。

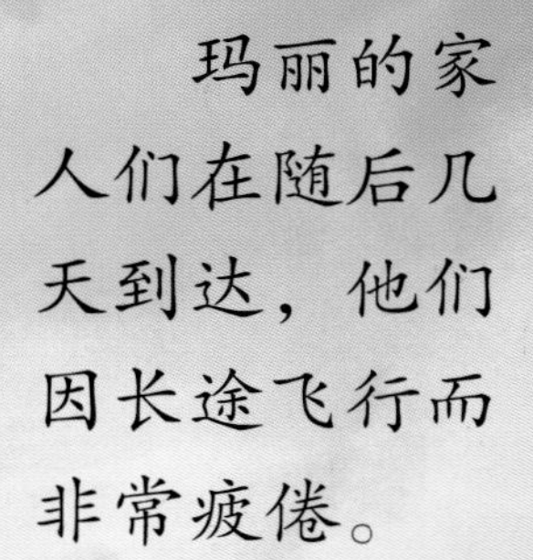

玛丽的家人们在随后几天到达，他们因长途飞行而非常疲倦。

玛丽向家人们问好，并向他们宣布：“以后我要和你们一起旅行！”玛丽在付完巨额账单后吸取了教训。

FACT OR FICTION?

阅读下面的句子，判断它们出自真实的部分还是虚构故事的部分。

1. 野鸭的骨头是中空的。

2. 野鸭吃牛排。

3. 野鸭会成群结队地飞。

4. 野鸭在它出生两个月左右之后才羽毛丰满。

答案

1 真实 2 虚构 3 真实 4 真实

FACT SPEED READ

Mallard ducklings are led to water by their mother as 10
soon as their soft, downy feathers dry off after hatching. 20

Young, down-covered ducklings cannot fly until 27
they are about two months old and their feathers have 37
grown in. 39

Ducks' bones are hollow, which makes their bodies 47
lighter. That makes it easier to fly long distances. 56

In the fall, mallards build up a store of fat on their 68
bodies. This provides them energy for the long flight 77
south. 78

Mallards eat wetland plants and grains such as wheat, 87
barley, and oats. They also eat some insects and shellfish. 97

Mallards spend the spring and summer in the 105
northern United States and Canada. They fly south for 114
the winter. 116

Ducks waddle when they walk. That's because their 124
legs are short and far apart. 130

Mallards fly in flocks rather than alone. A group of 140
birds is called a flock. 145

FICTION SPEED READ

Mally Mallard is swimming in the lake, minding her own business, when she notices a huge, gray object in the sky.

"Oh my, what is that?" she cries.

"It's called an airplane. People ride in it to go to faraway places," Daddy Drake says. "Soon you will learn to fly just like that airplane."

Mally is amazed! She wants to start right away! Mally's dad shows her what to do. Then Mally practices until she is flying high.

By the end of the day, Mally is very tired from all the flapping. She thinks that getting a ride on a plane sounds like a better idea than doing all this work!

The next day, Mally goes to the airport and jumps on a plane. As Mally sits down in her seat, the flight attendant asks, "Would you like lasagna or steak for lunch today?" Mally orders both!

After a few hours, the plane lands. Mally is 156
excited to get into the warm Texas air. On the way 167
off the plane, the attendant hands Mally an 175
envelope. In the envelope is a bill for $200! 184

Mally uses all of her money to pay the bill. She 195
is very disappointed and walks off the plane flat 204
broke. 205

Mally's family arrives several days later, very 212
tired from their long flight. When Mally greets her 221
family, she exclaims, "I will travel with you from 230
now on!" Mally has learned her lesson after paying 239
that huge bill! 242

GLOSSARY

bill. a list of costs or charges for items purchased

distance. the amount of space between two places

flat broke. completely without money

hollow. having an empty space inside

practice. to do over and over in order to learn a skill

store. a supply kept for future use

travel. to move from one place to another

wetland. a low, wet area of land such as a swamp or marsh

小收藏家瑞奇

鼠

［美］ 凯利·杜德娜　著
［美］ 尼娜·乔拉　绘　　王　妹　译

長春出版社
国　家　一　级　出　版　社
全国百佳图书出版单位

吉图字：07-2015-4474

图书在版编目（CIP）数据

我们身边的动物．小收藏家瑞奇·鼠 /（美）杜德娜著；（美）乔拉绘；王姝译．— 长春：长春出版社，2016.1

（动物故事）

ISBN 978-7-5445-4173-2

Ⅰ．①我… Ⅱ．①杜… ②乔… ③王… Ⅲ．①鼠科—儿童读物 Ⅳ．①Q95-49

中国版本图书馆 CIP 数据核字（2015）第 269706 号

著　　者　［美］凯利·杜德娜　　绘　　者　［美］尼娜·乔拉

译　　者　王　姝　　责任编辑　张　岚　刘　洋　单紫薇

長春出版社 出版　　吉林省吉广国际广告股份有限公司印制

（长春市建设街 1377 号）（邮编 130061　电话 88563443）　　长春出版社发行部发行

787 mm × 1092 mm　20 开　14.4 印张　108 千字

2016 年 1 月第 1 版　2016 年 1 月第 1 次印刷

定价：80.00 元（全 12 册）

www.cccbs.net

真实&虚构故事

这套充满奇思妙想的图书专为学龄前后的小读者设计。它以极为巧妙的方式，在引领小读者探索自然科学世界的同时，循序渐进地加强了小读者的阅读能力，并以自如穿梭于真实世界与虚构故事间的形式，启发小读者的思维。

FACT 真实：左侧页面展示了真实照片，以增强小读者的科学认知和对信息文本的理解。

FiCTiON 虚构故事：右侧页面通过一个有趣的叙事故事，配合奇思妙想的插图，来启发小读者无穷的想象力。

FACT OR FiCTiON? 真实和虚构故事的页面可以分别进行阅读，通过提问、预测、推断和总结的方式提高理解能力。也可以并排连续阅读，能够帮助小读者发掘和研究不同的写作风格。

真实还是虚构故事？通过有趣的测验，加强小读者对于何为真实何为虚构的理解。

书后特附英文原文和词汇表，并配有外教的纯正发音，以极为便捷的方式让小读者感受地道的英文。

FACT

棕鼠，或者叫挪威鼠，第一次从欧洲乘船到北美洲，大约是在 1776 年前后。

小老鼠瑞奇收藏了很多东西，他的小房间都塞满了。

“我是个收藏家，哈哈！看来，我需要更大一点儿的房间啦！”瑞奇骄傲地说。

FACT

美国的蒙大拿州和加拿大的阿尔伯达省的老鼠数量最少。

瑞奇先将他的宝贝整理了一部分，装进背包里。他找到了一个新鼠洞，离现在住的地方只隔了三条小巷子。

“这里看上去挺不错！”他说，“还能看到公园的风景呢！”

FACT

老鼠在下水道、建筑物深处、污水管和地道之类的地方筑巢。老鼠循规蹈矩，不喜欢改变。

瑞奇的新家里什么都没有。

“我要把这里装满！我想可以在公园里找到些东西！”

瑞奇说完，抓起他的背包，像一个狩猎者一样出发了。

FACT

老鼠可以将它们的身体骨架蜷缩起来，以挤进狭小的空间。它们甚至可以穿过像硬币大小的洞。

来到公园，瑞奇在花丛下发现了一个硬币。

旁边还有一个酒瓶的软木塞。

“太棒了！”瑞奇高兴得欢呼起来，“我可以将它们组合起来，做一个桌子！”

FACT

老鼠的黄色牙齿非常坚硬。它们能啃动铜管，甚至能嚼动水泥。

没走几步，一个闪闪发光的东西吸引了瑞奇的视线。

原来是一个玻璃瓶的金属盖。

“这个很适合给祖母的照片当相框！”瑞奇兴奋地说。

FACT

老鼠的嗅觉和听觉异常敏锐。但它们是色盲，而且视力很差。

瑞奇四处逛了一会儿，又发现了一个塑料瓶盖。“可以用它来装放在门前的植物。”

FACT

老鼠的游泳技能极佳。在紧急的情况下，它们能游很长一段距离，也可以在水下屏息 30 秒。

很快，瑞奇的背包就满了。他向公园外走时，路过一个喷水池，瑞奇毫不犹豫地跳下水，舒舒服服地游了起来。

“走了这么久，能在水里凉快一会儿真舒服！”瑞奇开心地说。

FACT

老鼠总是不断地咀嚼。它们的门牙每年会长 12 厘米。

瑞奇回到新家，倒出他找到的宝贝，然后将房间布置一新。

他环顾四周，一边毫不在意地咬着牙签，一边说：“我要收藏更多的东西，把这里填满！”

“明天，我要去寻找更多的宝贝！”

FACT OR FICTION?

阅读下面的句子，判断它们出自真实的部分还是虚构故事的部分。

1. 老鼠会用背包将物品带在身边。

2. 老鼠可以将它们的身体骨架蜷缩起来，以挤进狭小的空间。

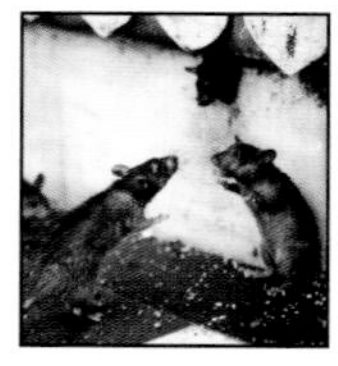

3. 老鼠会在它们的门前种植物。

4. 老鼠的门牙每年会长12厘米。

答案

1 虚构 2 真实 3 虚构 4 真实

FACT SPEED READ

The brown rat, or Norway rat, first came to North 10
America on ships from Europe around 1776. 17

Montana in the United States and Alberta in 25
Canada have the fewest rats. 30

Rats build nests under sidewalks and in buildings, 38
sewers, and subways. 41

Rats prefer routine to change. 46

Rats can collapse their skeletons to squeeze into 54
small spaces. They can go through a hole as small as a 66
quarter. 67

Rats' teeth are yellow and very hard. They can chew 77
through copper pipes and concrete. 82

Rats have very good senses of hearing and smell. 91
But they are color-blind and have very poor eyesight. 101

Rats are excellent swimmers. If they have to, they 110
can swim for a mile and can stay underwater for 30 121
seconds. 122

Rats gnaw constantly. Their front teeth grow about 130
five inches a year. 134

FICTION SPEED READ

Ricky Rat has collected so much stuff that his rat hole is full. "I'm such a pack rat. I need more space!" he declares.

Ricky packs a few things in his backpack. He finds a new hole three alleys over. "This looks like a good place," he says. "And it has a great view of the park."

Ricky's new hole is empty. He says, "I need to fill up this place! I bet I can find some things in the park." He grabs his backpack and sets off on a scavenger hunt.

At the park, Ricky finds a quarter under some flowers. The cork from a bottle is nearby. "This is great!" he exclaims. "I can put these together to make a table."

A few steps later, something shiny catches Ricky's eye. It is a metal bottle cap. He says, "This will make a great picture frame for my photo of Granny Rat."

Ricky wanders for a while before he sees a blue plastic bottle cap. "This can be a planter outside of my front door." 163 173 176

Ricky's backpack is full. On his way out of the park, he passes a fountain and jumps in for a swim. "Ooh, this feels good after all that walking around," he says. 186 197 206 208

Ricky returns to his new rat hole, unpacks all of his new things, and sets everything up. He looks around the room. "I could pack a lot more in here," he says, absently chewing on a toothpick. "I'll have to look for more stuff tomorrow!" 218 227 238 247 253

GLOSSARY

absently. not paying attention

color-blind. partially or totally unable to tell one color from another

cork. a plug for a bottle or jug

gnaw. to bite or chew on with the teeth

pack rat. 1) a type of wood rat known for hoarding food and other things 2) someone who collects and keeps a lot of unneeded things

sewer. an underground passage used to carry away waste

subway. an underground passage for people or trains

动物故事 · 我们身边的动物

罗比的长耳朵

［美］ 凯利·杜德娜 著

［美］ C.A. 诺本斯 绘 王 妹 译

国 家 一 级 出 版 社

全国百佳图书出版单位

吉图字：07-2015-4474

图书在版编目（CIP）数据

我们身边的动物．罗比的长耳朵·兔子／（美）杜德娜著；（美）诺本斯绘；王妹译．—长春：长春出版社，2016.1
（动物故事）
ISBN 978-7-5445-4173-2

Ⅰ．①我… Ⅱ．①杜… ②诺… ③王… Ⅲ．①兔科—儿童读物 Ⅳ．①Q95-49

中国版本图书馆 CIP 数据核字（2015）第 269707 号

著　者［美］凯利·杜德娜　　绘　者［美］C.A. 诺本斯
译　者 王　妹　　责任编辑 张　岚　刘　洋　单紫薇

長春出版社 出版　　吉林省吉广国际广告股份有限公司印制
（长春市建设街 1377 号）（邮编 130061　电话 88563443）　　长春出版社发行部发行
787 mm×1092 mm　20 开　14.4 印张　108 千字
2016 年 1 月第 1 版　2016 年 1 月第 1 次印刷
定价：80.00 元（全 12 册）
www.cccbs.net

真实&虚构故事

这套充满奇思妙想的图书专为学龄前后的小读者设计。它以极为巧妙的方式，在引领小读者探索自然科学世界的同时，循序渐进地加强了小读者的阅读能力，并以自如穿梭于真实世界与虚构故事间的形式，启发小读者的思维。

FACT 真实：左侧页面展示了真实照片，以增强小读者的科学认知和对信息文本的理解。

FICTION 虚构故事：右侧页面通过一个有趣的叙事故事，配合奇思妙想的插图，来启发小读者无穷的想象力。

FACT OR FICTION? 真实和虚构故事的页面可以分别进行阅读，通过提问、预测、推断和总结的方式提高理解能力。也可以并排连续阅读，能够帮助小读者发掘和研究不同的写作风格。

真实还是虚构故事？通过有趣的测验，加强小读者对于何为真实何为虚构的理解。

书后特附英文原文和词汇表，并配有外教的纯正发音，以极为便捷的方式让小读者感受地道的英文。

FACT

兔子经常会在清晨和傍晚时出来活动。白天的时候，它们会在隐蔽的洞穴里休息。

小兔子罗比想看电视。可是电视的画面有些模糊，罗比将兔耳形电视天线鼓捣了一阵也无济于事。“没有用呀！”他说，“我还是不看了，去外面玩吧！”

FACT

野兔吃青草和植物的细茎，也会吃一些蔬菜和小花。

罗比在出门前经过厨房。“先吃点东西吧！”他想。

他吃了点葡萄干，又拿出个橘子。

FACT

野兔生活在森林的边缘、田地的周围，或者其他草丛密集的地方。

罗比遇到了他的折耳兔表兄妹们，贝蒂、比特西和鲍伯。

他们正在森林边缘玩耍。

“嘿，罗比！”他们向罗比喊道，“我们来玩捉迷藏怎么样？”

FACT

兔子的听力非常棒，大多数兔子的耳朵很长。有些家养的兔子，耳朵低垂或者完全是折耳。

“你的耳朵可以垂下来，这可真幸运。我的耳朵总是竖着，所以不管我藏到哪儿，总会立刻就被发现。”罗比抱怨道，“但这个游戏还是很好玩！”

比特西开始数数，“1、2、3……”

FACT

大多数野兔的共同点是它们都属于白尾灰兔。白尾灰兔的尾巴呈棕白色，毛茸茸的。

罗比躲到了园子里。

“怎样才能藏起我的长耳朵呢？”罗比犯难。

稻草人听到了他的自言自语。

“你可以戴我的帽子。”稻草人建议道。

罗比戴上了帽子。“太管用啦！”

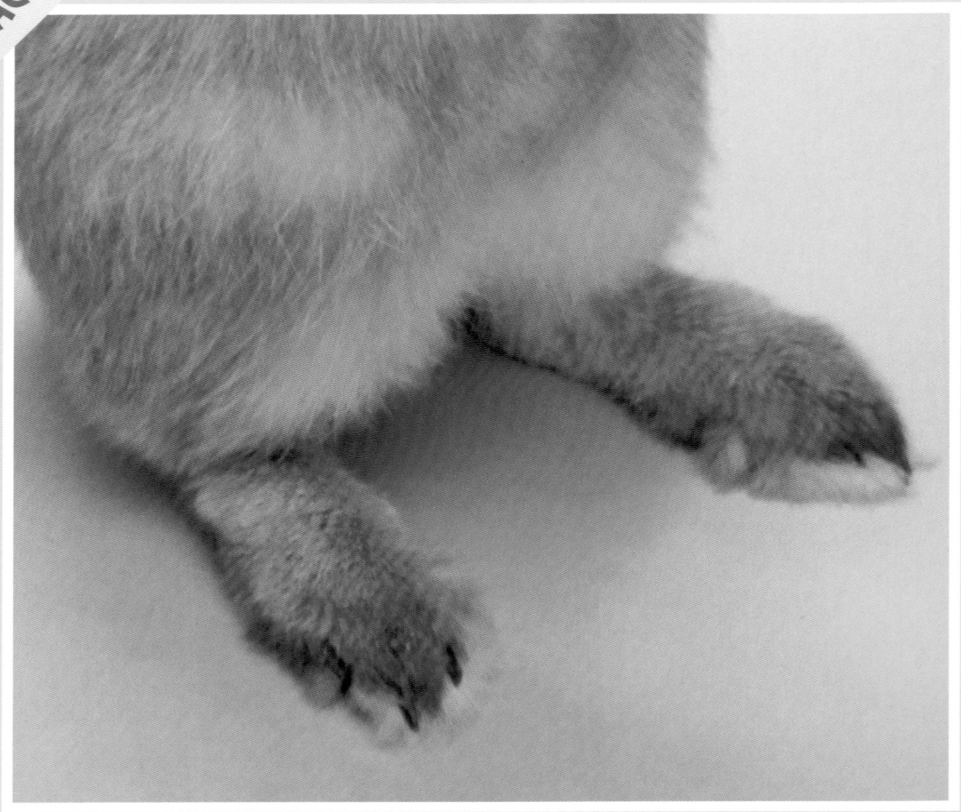

FACT

兔子的后脚很长。它们会用后脚捶地，来向同伴通报危险将近。

罗比安静地躲着，可是没人发现他。就在他想跑回去的时候，罗比听到他表哥的捶足声。

罗比发现一只大狗正向他走来。

在危险来临的时候，兔子会尽力逃走或在原地屏息不动。兔子能够在很长一段时间内保持一动不动。

“哦！不！这只大狗会看到我的耳朵！”他随即想到，他戴着稻草人的帽子呢！

罗比一动不动，直到大狗离开。

FACT

兔子的眼睛长在脑袋的两侧。它们的眼睛能够看到任何一个方向。

在确保安全后，罗比跳回到表兄身边。鲍伯问："你看见大狗了吗？"

罗比咧嘴一笑回答道："看到啦！但是我的耳朵藏起来啦！"

"那是什么意思？"贝蒂歪着头问。

"稻草人把他的帽子给我了。那只狗就找不到我了，连你们也找不到我！我们来玩抓人吧！"罗比说完大笑起来。

FACT OR FICTION?

阅读下面的句子，判断它们出自真实的部分还是虚构故事的部分。

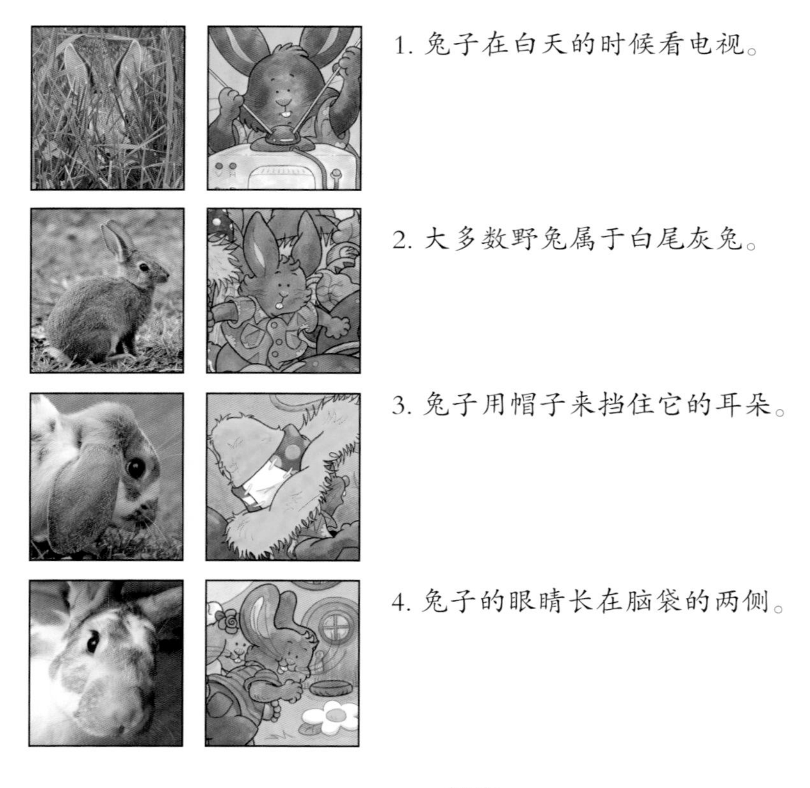

1. 兔子在白天的时候看电视。

2. 大多数野兔属于白尾灰兔。

3. 兔子用帽子来挡住它的耳朵。

4. 兔子的眼睛长在脑袋的两侧。

答案

1 虚构 2 真实 3 虚构 4 真实

FACT SPEED READ

Rabbits are most active at dawn and dusk. During 9
the day, they rest in hidden spots. 16

Wild rabbits eat grass and small twigs. They will also 26
eat garden vegetables and flowers. 31

Wild rabbits live at the edges of forests and fields or 42
any other place with brushy cover. 48

Rabbits have excellent hearing, and most have long 56
ears. Some domestic rabbits have ears that hang 64
down, or lop ears. 68

The most common wild rabbit is the cottontail 76
rabbit. Cottontails are grayish-brown and have white, 84
fluffy tails. 86

Rabbits have long hind feet. They thump their hind 95
feet to warn other rabbits of danger. 102

When there is danger, a rabbit will run away or 112
freeze where it sits. Rabbits can sit completely still for a 123
long time. 125

A rabbit's eyes are on the sides of its head. Rabbits 136
can see in all directions. 141

FICTION SPEED READ

Robbie Rabbit wants to watch TV. The picture is fuzzy, so Robbie fiddles with the rabbit ears. "That didn't help," he says. "I'll go outside and play instead."

Robbie passes through the kitchen on the way outside. "I'll have a snack first," he thinks to himself. He nibbles on raisins and an orange.

Robbie meets his lop-eared cousins, Betty, Bitsy, and Bob. They are playing at the edge of the woods. "Hey, Robbie!" they shout. "How about a game of hide-and-seek?"

"You're lucky that your ears hang down. Mine stick up, so you always find me right away," Robbie complains. "But it's still fun to play."

Bitsy starts counting, "One, two, three ..."

Robbie scampers to the garden. "How can I hide my ears?" he wonders.

The scarecrow hears him. "Wear my hat," he offers.

Robbie puts it on. “It works!” he exclaims. 145

Robbie waits, but no one finds him. He is just 155
about to run back when he hears one of his cousins 166
thumping. Robbie sees a big dog walking straight 174
toward him! 176

Robbie thinks, “Oh, no! That dog will see my 185
ears!” Then he remembers that he’s wearing the 193
scarecrow’s hat. He stays very still until the dog is 203
gone. 204

When it is safe, Robbie hops back to his cousins. 214
Bob asks, “Did you see the dog?” 221

Robbie grins and replies, “Yes, but my rabbit ears 230
were down.” 232

“What do you mean?” Betty wonders. 238

“The scarecrow gave me his hat. The dog couldn’t 247
find me and neither could you! Let’s play tag!” 256
Robbie says with a laugh. 261

GLOSSARY

dawn. the time of day when the sky grows lighter and the sun rises

dusk. the time of day when the sky grows darker and the sun sets

freeze. to become completely motionless

lop-eared. having ears that droop down

rabbit ears. an indoor TV antenna with two rods that extend up and out from a common point to form a V shape

scarecrow. a dummy, usually human-shaped, that is set up in a field or garden to scare crows and other birds away from the crops

约德尔音乐比赛

［美］凯利·杜德娜　著
［美］安妮·哈伯斯通　绘　　王　妹　译

国　家　一　级　出　版　社
全国百佳图书出版单位

吉图字：07-2015-4474

图书在版编目（CIP）数据

我们身边的动物．约德尔音乐比赛·狗 /（美）杜德娜著；（美）哈伯斯通绘；王妹译．— 长春：长春出版社，2016.1
（动物故事）
ISBN 978-7-5445-4173-2

Ⅰ．①我… Ⅱ．①杜… ②哈… ③王… Ⅲ．①犬－儿童读物 Ⅳ．①Q95-49

中国版本图书馆 CIP 数据核字（2015）第 269755 号

著　　者 ［美］凯利·杜德娜　　绘　　者 ［美］安妮·哈伯斯通
译　　者 王　妹　　责任编辑 张　岚　刘　洋　单紫薇

長春出版社 出版　　吉林省吉广国际广告股份有限公司印制
（长春市建设街 1377 号）（邮编 130061　电话 88563443）　　长春出版社发行部发行
787 mm×1092 mm　20 开　14.4 印张　108 千字
2016 年 1 月第 1 版　2016 年 1 月第 1 次印刷
定价：80.00 元（全 12 册）
www.cccbs.net

真实&虚构故事

这套充满奇思妙想的图书专为学龄前后的小读者设计。它以极为巧妙的方式，在引领小读者探索自然科学世界的同时，循序渐进地加强了小读者的阅读能力，并以自如穿梭于真实世界与虚构故事间的形式，启发小读者的思维。

FACT 真实：左侧页面展示了真实照片，以增强小读者的科学认知和对信息文本的理解。

FICTION 虚构故事：右侧页面通过一个有趣的叙事故事，配合奇思妙想的插图，来启发小读者无穷的想象力。

FACT OR FICTION? 真实和虚构故事的页面可以分别进行阅读，通过提问、预测、推断和总结的方式提高理解能力。也可以并排连续阅读，能够帮助小读者发掘和研究不同的写作风格。

真实还是虚构故事？通过有趣的测验，加强小读者对于何为真实何为虚构的理解。

书后特附英文原文和词汇表，并配有外教的纯正发音，以极为便捷的方式让小读者感受地道的英文。

FACT

狗通过喘息散热。它们吸入新鲜、凉爽的空气，然后呼出热气，并将体内额外的热量一起排出。

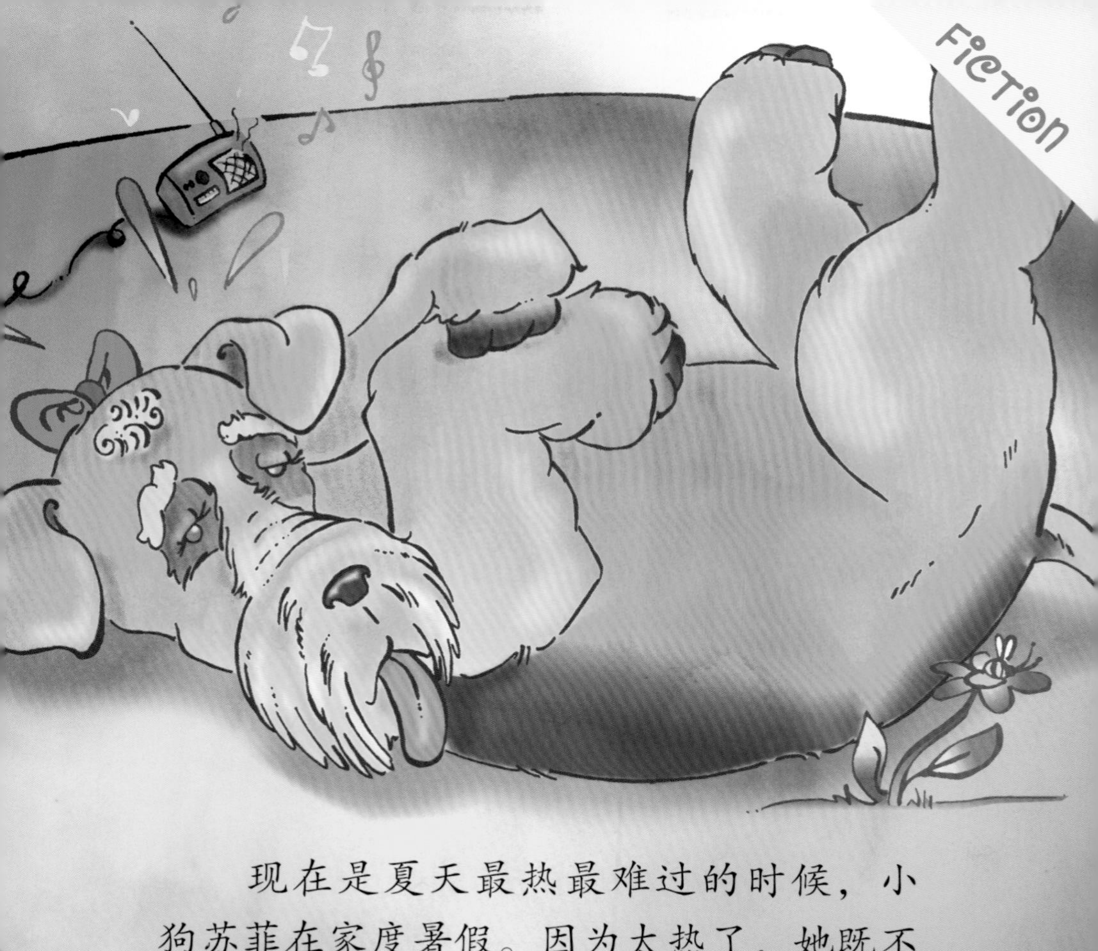

现在是夏天最热最难过的时候，小狗苏菲在家度暑假。因为太热了，她既不想玩捡木棍儿，也不想玩埋骨头。苏菲只想听着收音机，躺在地上不停地喘气。

FACT

狗能听到的声音距离，比人能听到的远四倍。

苏菲在收音机里听到了一个约德尔音乐比赛的广告。比赛的获胜者将获得去瑞士山上的约德尔音乐学校参观的机会。

“我可以用真声和假声转换的唱法。”苏菲心里想，“路易斯和我都可以去参加比赛！”

狗通过吠声、号叫、咕哝声、哀诉声、狂叫，相互交流。

苏菲打电话给她最好的朋友路易斯，告诉她比赛的消息。

“别忘带上你的手风琴，我们一起练习表演。”苏菲对路易斯说。

FACT

很多狗会跟着汽笛叫。专家认为，狗可能以为汽笛声是同伴发出的信号。

正在苏菲和路易斯一起练习的时候，一辆消防车急速驶过。隔壁的狗们都跟着“嗷喔喔喔喔！”地吼了起来。苏菲信心满满地说：“我们的声音可比他们棒多啦！”

FACT

狗只能看出少量的颜色，而且它们看到的画面没有人看到的那样生动。

“我们需要一些滑稽的衣服。”苏菲说。

她和路易斯一起去了服装道具店。

“快看这条皮短裤！”苏菲兴奋地对路易斯说。

“嗯，真不错！”路易斯赞同道。

FACT

狗的体重在 2 千克至 90 千克之间。

比赛那天到了。

苏菲唱了一首《哦嘞依哦》。路易斯演奏手风琴。他们比其他的选手唱得都好。苏菲和路易斯顺利地拿到了大奖。

FACT

狗对细致的事物观察不详，却对运动的事物非常敏感。

苏菲和路易斯透过飞机窗俯视群山。飞机在瑞士山着陆之前，他们几乎抑制不了激动、兴奋的心情。

FACT

不论是和人类，还是与其他品种的狗一起玩耍，狗总是很活跃。

苏菲和路易斯来到约德尔音乐学校后，很快加入到一群比他们早到的擅长约德尔唱法的伙伴当中。

“哦，路易斯！”苏菲愉快地跟她的朋友说，“这里真是非常有趣！”

FACT OR FICTION?

阅读下面的句子，判断它们出自真实的部分还是虚构故事的部分。

1. 狗能听到的声音距离，比人能听到的远四倍。

2. 狗会演奏手风琴。

3. 狗只能看出少量的颜色。

4. 狗穿皮短裤。

答案

1真实2虚构3真实4虚构

FACT SPEED READ

Dogs pant to cool off. They inhale fresh, cool air and exhale warm air that carries away extra body heat. 10 20

Dogs can hear sounds up to four times farther away than humans can. 30 33

Dogs howl, growl, grunt, whine, and bark to communicate with each other. 41 45

Many dogs howl along with sirens. Experts believe that dogs think sirens sound like other dogs howling. 53 62

Dogs can see a little color, but much less vividly than humans do. 72 75

Dogs range in weight from under five pounds to over 200 pounds. 84 87

Dogs see little detail but are very sensitive to movement. 96 97

Dogs thrive on social play, whether it's with their human family or other dogs. 106 111

FICTION SPEED READ

It's the dog days of summer, and Sophie Dog is on vacation from school. It's too hot to fetch any sticks or bury any bones. Instead, Sophie just lies around panting and listening to the radio.

Sophie hears an ad on the radio for a yodeling contest. The grand prize is a trip to a yodeling school in the mountains of Switzerland. "I can yodel," Sophie thinks to herself. "Louise and I should enter that contest!"

Sophie phones her best friend, Louise, and tells her about the contest. "Bring your accordion over, and we will practice our act," Sophie says to Louise.

As Sophie and Louise practice, the fire trucks leave from the station next door. The other dogs in the neighborhood howl, "Owooo, owooo, owooooooo!"

Sophie declares, "We sound so much better than that!"

"We need something fun to wear," Sophie says. 143
She and Louise go to a costume shop. "Look at 153
these green lederhosen!" Sophie exclaims. 158

"They're perfect!" Louise agrees. 162

The day of the contest arrives. Sophie sings, 170
"Yodel-ay-hee-hoo." Louise plays along on her 179
accordion. They sound better than all of the other 188
dogs. They easily win the grand prize. 195

Sophie and Louise look out the airplane windows 203
at the mountains. They can barely contain their 211
excitement as the airplane lands in Switzerland. 218

When Sophie and Louise arrive at the yodeling 226
school, they join the crowd of dog yodelers who are 236
already there. "Oh Louise!" Sophie exclaims to her 244
friend. "This is going to be so much fun!" 253

GLOSSARY

accordion. a handheld musical instrument that produces sound when air is forced through it by squeezing the two ends together

dog days. the period of hot weather between July and September

exhale. to breathe out

inhale. to breathe in

lederhosen. leather shorts with suspenders that are traditionally worn in the European Alps

yodel. to sing in a style that changes rapidly between normal tones and high, false tones

追踪库巴团

猫

[美] 帕姆 · 斯凯内曼 著
[美] 尼娜 · 乔拉 绘　王 妹 译

国 家 一 级 出 版 社
全国百佳图书出版单位

吉图字：07-2015-4474

图书在版编目（CIP）数据

我们身边的动物．追踪库巴团·猫 /（美）斯凯内曼著；（美）乔拉绘；王姝译．— 长春：长春出版社，2016.1
（动物故事）
ISBN 978-7-5445-4173-2

Ⅰ．①我… Ⅱ．①斯… ②乔… ③王… Ⅲ．①猫—儿童读物 Ⅳ．①Q95-49

中国版本图书馆CIP数据核字（2015）第269759号

著　　者 [美]帕姆·斯凯内曼　　绘　　者 [美]尼娜·乔拉
译　　者 王　姝　　责任编辑 张　岚　刘　洋　单紫薇

長春出版社 出版　　吉林省吉广国际广告股份有限公司印制
（长春市建设街1377号）（邮编130061　电话88563443）　　长春出版社发行部发行
787㎜×1092㎜　20开　14.4印张　108千字　　定价：80.00元（全12册）
2016年1月第1版　2016年1月第1次印刷

www.cccbs.net

真实&虚构故事

这套充满奇思妙想的图书专为学龄前后的小读者设计。它以极为巧妙的方式，在引领小读者探索自然科学世界的同时，循序渐进地加强了小读者的阅读能力，并以自如穿梭于真实世界与虚构故事间的形式，启发小读者的思维。

FACT 真实：左侧页面展示了真实照片，以增强小读者的科学认知和对信息文本的理解。

FICTION 虚构故事：右侧页面通过一个有趣的叙事故事，配合奇思妙想的插图，来启发小读者无穷的想象力。

FACT OR FICTION? 真实和虚构故事的页面可以分别进行阅读，通过提问、预测、推断和总结的方式提高理解能力。也可以并排连续阅读，能够帮助小读者发掘和研究不同的写作风格。

真实还是虚构故事？通过有趣的测验，加强小读者对于何为真实何为虚构的理解。

书后特附英文原文和词汇表，并配有外教的纯正发音，以极为便捷的方式让小读者感受地道的英文。

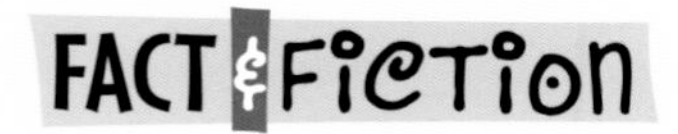

FACT

猫儿们会花很多时间打扮自己。它们舔舐身上的皮毛，以保持干净，这样也能在热的时候，给自己降温。

休吉是一只非常特别的猫。他的皮毛可以变色，这可以帮助他混入周围的环境而不被发现。这个特长对作为波斯威尔猎犬侦查队队长的他来说，尤其有帮助。

FACT

猫的听力非常敏锐。它们会用耳朵定位声音的来源。

Fiction

今天，休吉和他的得力助手安琪正在追查企图占领波斯威尔的一群库巴团猎犬。

FACT

它们喝水的时候用舌头下面舔，而不用舌头上面。

休吉和安琪来到咖啡店，各自要了杯咖啡。他们来这里监视一只黑色的拉布拉多犬。“瞧，他点了那么多咖啡。”

休吉说着，抿了口咖啡。“我猜他就是库巴团组织的成员，走吧！跟着他！”

FACT

猫儿们用尾巴相互交流。当它们被惹恼，表现出攻击的姿态时，它们会将尾巴左右摆动。

休吉和安琪跟踪那只拉布拉多犬来到沼泽地。休吉说："我混进去，看看他要去哪儿。"

"好的，老大！我在这儿等你，我会开着对讲机的。"安琪说道。他一想到可以握枪抓捕猎犬，就兴奋得尾巴直摇。

FACT

猫儿们的胡子与它们的身体同宽。它们用胡子来估量身体是否可以穿过狭窄的通道。

休吉的皮毛变成了叶子的颜色。他穿过树木间的缝隙，紧随着目标。

库巴团由一只非常有名的猎犬领导。他们在蒲草当中的空地聚集。

猫儿们每天要睡 16 到 18 个小时。即使在睡觉的时候，它们也对危险十分警觉。

休吉谨慎地伏在空地边缘的蒲草中。现在，他的毛发变得和蒲草的颜色一样啦。他的身体前倾，仔细听猎犬们的动静。

猎犬的首领说："明天，我们要开始行动！就在那群懒猫打盹儿的时候！"

FACT

猫儿们的嗅觉灵敏度要比人类高很多倍。

休吉小声用对讲机告诉安琪："我已经窃听到他们的计划了。现在你打电话通知总部，请总部立刻发消息给远程超级猫指挥中心。"

FACT

猫喜欢高的地方。它们可以跳到是它们身长6倍的地方。

“库巴团的猎犬们对抗不了远程超级猫指挥中心的木棍战略！这样，他们就会一路追踪那些木棍，从而远离我们的波斯威尔。”

休吉和安琪高兴地相互击掌。

“又完成了一项任务！”他们心满意足地说。

FACT OR FICTION?

阅读下面的句子，判断它们出自真实的部分还是虚构故事的部分。

1. 猫儿们的胡子与身体同宽。

2. 猫儿们喜欢较高的地方。

3. 猫儿们能变色。

4. 猫儿们喝咖啡。

答案

1 真实 2 真实 3 虚构 4 虚构

FACT SPEED READ

Cats spend a lot of time grooming. Licking their fur 10
keeps them clean. It also cools them when they are 20
hot. 21

Cats hear very well. They rotate their ears to locate 31
the source of a sound. 36

When cats lap up liquids, they use the underside of 46
their tongues instead of the top. 52

Cats use their tails to communicate. When cats are 61
annoyed, or about to attack, they wave their tails from 71
side to side. 74

The span of a cat's whiskers is about as wide as its 86
body. Cats use their whiskers to see if they can fit 97
through openings. 99

Cats sleep 16 to 18 hours a day. But they stay alert 111
to danger even when they are asleep. 118

Cats have a sense of smell many times stronger than 128
that of humans. 131

Cats love high places. Cats can jump up to six times 142
as high as they are long. 148

FICTION SPEED READ

Silky is a very special cat. He can change the color of his fur to blend in to any surroundings. Being able to alter his appearance is very helpful in his job as the head detective of the Purrsville Pooch Patrol.

Today Silky and his able assistant, Angel, are checking out tips that the Bowser Bunch dog pack is planning to take over Purrsville.

Silky and Angel stop at the café for a cup of coffee. There they see a black Lab dog. "Look how many extra cups of coffee he's buying," Silky says, sipping his coffee. "I bet he's a member of the Bowser Bunch. Let's go!"

Silky and Angel tail the dog to the swamp. Silky says, "I'm going to blend in and see where he goes."

"I'll stay here, boss, and listen to the police radio," Angel says. Her tail twitches in excitement at the thought of nabbing the gang.

Silky's fur takes on a leafy pattern. He crawls through a hollow log as he follows the dog. The Bowser Bunch, led by the notorious Big Dog, is gathered in a clearing in the cattails.

Silky creeps silently to the edge of the clearing. By now he has taken on the appearance of the cattails. He leans forward to listen.

"We'll have to make our move tomorrow while those lazy cats are catnapping," Big Dog says.

Silky whispers into the radio. He says to Angel, "I've sniffed out their plan. I want you to call headquarters. Ask them to send over the remote control Super Stick Cat-a-Pult right away."

The Bowser Bunch dogs can't resist chasing the sticks thrown by the remote control Super Stick Cat-a-Pult. They chase them right out of Purrsville. Silky and Angel high-five each other. "It's another job well done!" they purr with satisfaction.

GLOSSARY

alert. watchful and aware of what is happening

alter. to change

communicate. to share ideas, information, or feelings

groom. to clean oneself and take care of one's appearance

nab. to arrest or capture

notorious. having a widely known, bad reputation

rotate. to turn on or around a center

tail. 1) a part of an animal's body that sticks out from its rear end 2) to follow and watch someone

朵提学做冰激凌

奶牛

[美] 帕姆·斯凯内曼 著
[美] 安妮·哈伯斯通 绘 王 妹 译

長春出版社
国 家 一 级 出 版 社
全国百佳图书出版单位

吉图字：07-2015-4474

图书在版编目（CIP）数据

我们身边的动物．朵提学做冰激凌·奶牛 /（美）斯凯内曼著；（美）哈伯斯通绘；王妹译．— 长春：长春出版社，2016.1

（动物故事）

ISBN 978-7-5445-4173-2

Ⅰ．①我… Ⅱ．①斯… ②哈… ③王… Ⅲ．①乳牛—儿童读物 Ⅳ．①Q95-49

中国版本图书馆CIP数据核字（2015）第269767号

著　者［美］帕姆·斯凯内曼　　绘　者［美］安妮·哈伯斯通

译　者　王　妹　　责任编辑　张　岚　刘　洋　单紫薇

長春出版社 出版　　吉林省吉广国际广告股份有限公司印制

（长春市建设街1377号）（邮编130061　电话88563443）　　长春出版社发行部发行

787 mm×1092 mm　20开　14.4印张　108千字

2016年1月第1版　2016年1月第1次印刷

定价：80.00元（全12册）

www.cccbs.net

真实&虚构故事

这套充满奇思妙想的图书专为学龄前后的小读者设计。它以极为巧妙的方式，在引领小读者探索自然科学世界的同时，循序渐进地加强了小读者的阅读能力，并以自如穿梭于真实世界与虚构故事间的形式，启发小读者的思维。

FACT 真实：左侧页面展示了真实照片，以增强小读者的科学认知和对信息文本的理解。

FICTION 虚构故事：右侧页面通过一个有趣的叙事故事，配合奇思妙想的插图，来启发小读者无穷的想象力。

FACT OR FICTION? 真实和虚构故事的页面可以分别进行阅读，通过提问、预测、推断和总结的方式提高理解能力。也可以并排连续阅读，能够帮助小读者发掘和研究不同的写作风格。

真实还是虚构故事？通过有趣的测验，加强小读者对于何为真实何为虚构的理解。

书后特附英文原文和词汇表，并配有外教的纯正发音，以极为便捷的方式让小读者感受地道的英文。

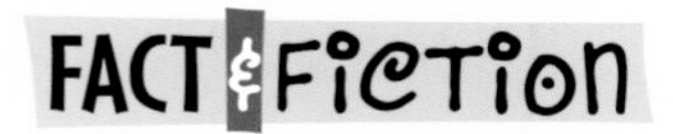

FACT

雌性奶牛在幼年时被称为小母牛。直到它们生下幼崽，才会被叫作奶牛。

奥利瑞夫人是一头很出名的奶牛。她是摩奥小镇上唯一一家冰激凌店——“冰爽一下”的主人。

奶牛的种类有很多。最常见的奶牛是黑白花的荷兰奶牛。

远近的奶牛和小母牛每个周六都会来这里，用美味的冰激凌犒劳自己。这已成为小镇的传统。

FACT

每头荷兰奶牛身上的斑点都不相同，像雪花一样，没有两头奶牛身上的标记是一样的。

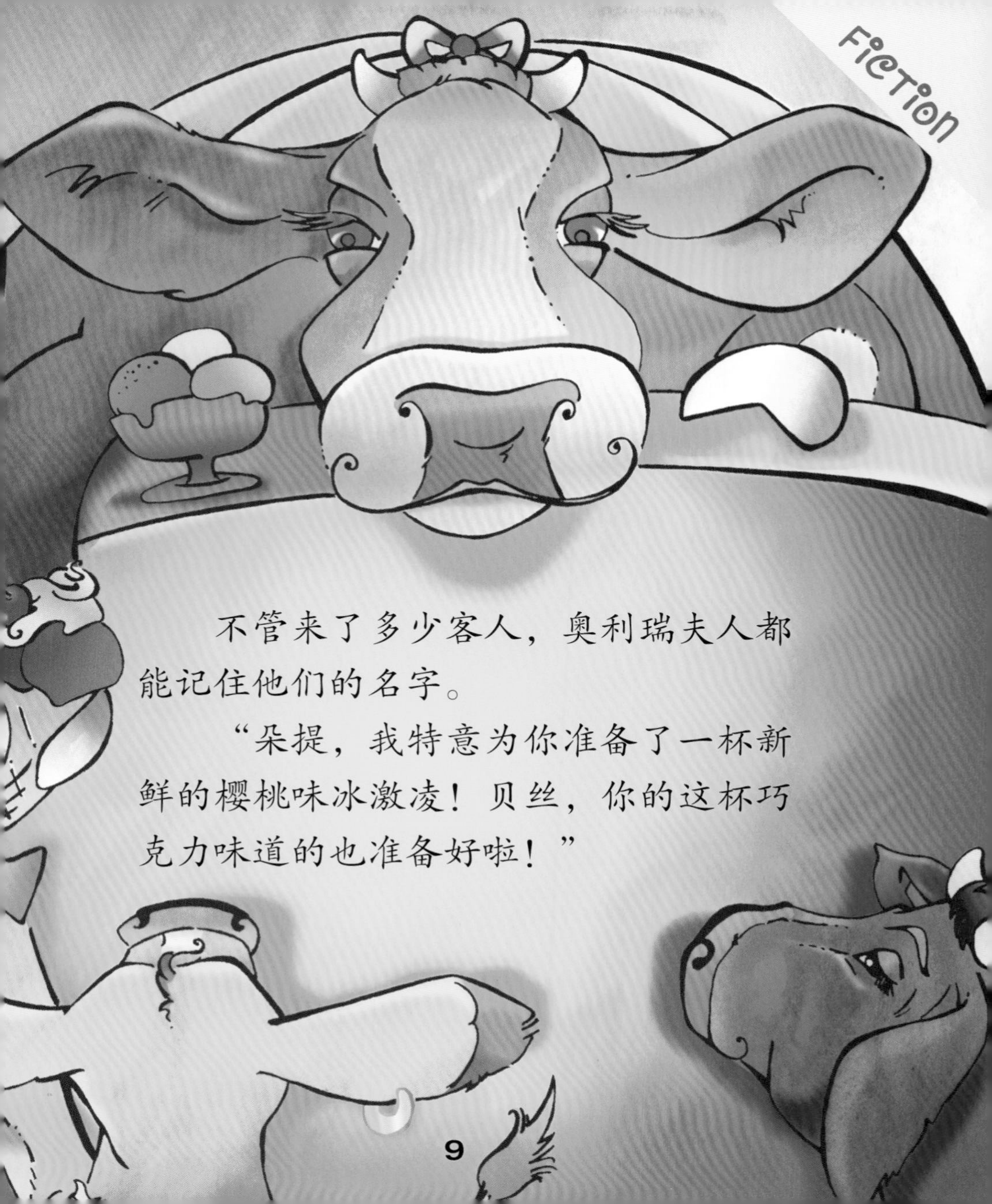

不管来了多少客人，奥利瑞夫人都能记住他们的名字。

“朵提，我特意为你准备了一杯新鲜的樱桃味冰激凌！贝丝，你的这杯巧克力味道的也准备好啦！”

FACT

奶牛每天会花上 8 个小时的时间进食。它们每天喝 90 升至 180 升的水。

朵提非常喜欢奥利瑞夫人，她通常会在其他客人离开后，再要一杯苏打水，多待上一会儿。朵提问：“奥利瑞夫人，您可以教我做冰激凌吗？如何能像您做得那么棒？”

FACT

奶牛没有上面的门牙。它们吃草的时候，是用舌头来回搅动或是用下牙咬碎。

“哦，亲爱的朵提，我很乐意教你！下周六的摩奥镇庆典，我很需要你来帮忙一起准备。”奥利瑞夫人说，“来吧，到柜台后面来，戴上围裙。我们先来准备桃子！”

FACT

牛会将它们的食物整个吞掉。这些食物首先进入它们四个胃中的第一个——瘤胃（另三个胃分别是网胃、瓣胃、皱胃）。牛会反刍食物，将已经吞下的食物逆呕、再嚼。

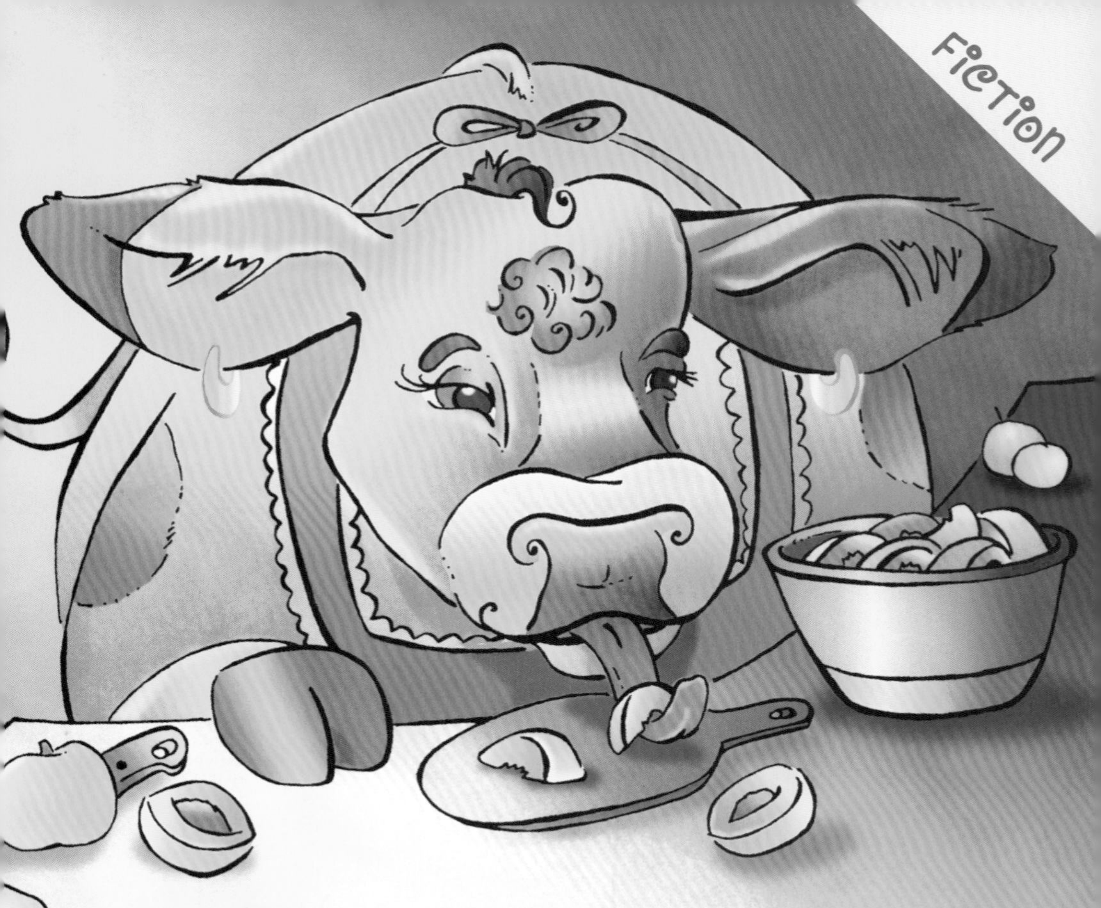

朵提仔细地清洗桃子，然后认真地将它们切成小块儿。在将切好的桃子放进大碗之前，她已经吃掉了不少。奥利瑞夫人笑道：“朵提，为做冰激凌多留点桃子吧！”

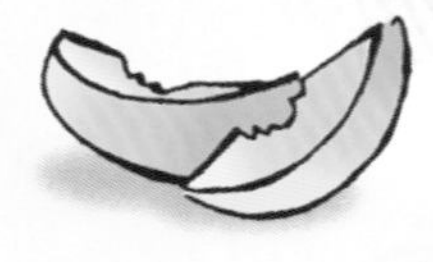

FACT

如果奶牛的食物中未能含有充足的盐分，会导致它们失去食欲，产奶量变少。所以，放奶牛的地方通常会有一块盐渍地供它们舔舐。

奥利瑞夫人在容器中加入糖、奶油和鸡蛋。然后，她又在旁边准备盐和冰。朵提一刻不停地看着冰激凌制作机，直到她筋疲力尽。“这个工作真难啊！”朵提感慨道：“谢谢您教我如何做冰激凌！”

FACT

如果你的头顶有一缕头发和周围头发生长方向不同，那就是发旋儿，英文用 cowlick 一词表示。也许，就是因为看起来像是被牛舔过的样子。

摩奥小镇庆典的那天天气很好，阳光充足、温暖。奥利瑞夫人看到了朵提，给了她一大碗冰激凌。吃光这碗冰激凌后，朵提高兴地说：“哇！这真是我吃过的最棒的冰激凌啦！”

FACT OR FICTION?

阅读下面的句子，判断它们出自真实的部分还是虚构故事的部分。

1. 奶牛穿围裙。

2. 奶牛每天花 8 个小时进食。

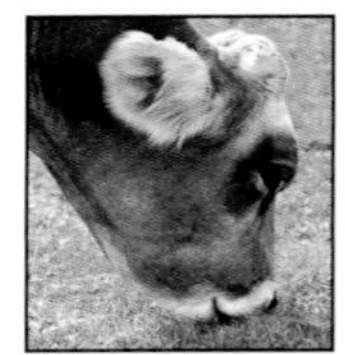
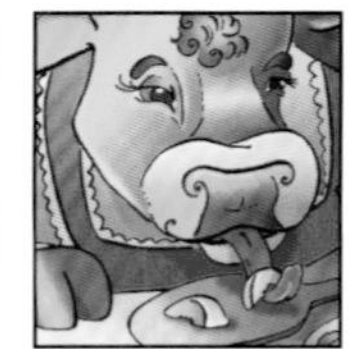

3. 奶牛会将食物整个吞下。

4. 奶牛在冰激凌店工作。

答案

1 虚构 2 真实 3 真实 4 虚构

FACT SPEED READ

Young female cattle are called heifers. After female 8
cattle have had calves, they are called cows. 16

There are many types of cows. The most common 25
dairy cow is the black-and-white Holstein. 33

The spots on Holstein cows are unique to each cow. 43
Like snowflakes, no two cows' markings are alike. 51

Dairy cows spend up to eight hours a day eating. 61
They drink 25 to 50 gallons of water a day. 71

Cows do not have top front teeth. Cows grab grass 81
by twisting it around their tongues and pulling it or 91
breaking it with their bottom front teeth. 98

Cows swallow their food whole. It goes into the first 108
of four compartments in their stomachs. Cows 115
regurgitate the resulting cud and chew it. 122

A dairy cow that doesn't get enough salt in its diet 133
may lose its appetite and produce less milk. Cows are 143
usually provided with a salt lick. 149

A tuft of hair that won't lie flat is called a cowlick. 161
Perhaps that's because it looks like hair a cow has licked. 172

FICTION SPEED READ

Mrs. O'Leary is a proud cow. She is the owner of Moo Town's only ice-cream shop, Cool Licks.

Calves and cows from near and far gather here each Saturday for their weekly ice-cream treat. It's a Cattle County tradition.

No matter how many come, Mrs. O'Leary knows each of them by name. "Dottie, I've got a fresh batch of cherry ice cream just for you! Bessie, your bowl of chocolate is almost ready."

Dottie admires Mrs. O'Leary so much that she likes to stay and have a soda after the others have gone. Dottie asks, "Mrs. O'Leary, will you teach me how to make ice cream as wonderful as yours?"

"Dottie, I'd love to teach you! I'll need the help to get ready for the Moo Town Festival next Saturday," Mrs. O'Leary says. "Come behind the counter and put on an apron. Let's start with those peaches."

Dottie washes the peaches and carefully chops 156
them into pieces. She tastes quite a few as she puts 167
them in the bowl. Mrs. O'Leary says, "Dottie, save 176
some for the ice cream!" 181

Mrs. O'Leary adds the sugar, cream, and eggs. 189
Then she packs salt and ice around the container. 198
Dottie cranks the ice-cream maker until she's worn 207
out. "This is hard work!" Dottie exclaims. "Thanks 215
for teaching me how to make ice cream." 223

The day of the Moo Town Festival is warm and 233
sunny. Mrs. O'Leary sees Dottie and gives her a big 243
bowl of ice cream. 247

After she licks the bowl clean, Dottie says, "Wow! 256
This is the best ice cream I've ever had!" 265

GLOSSARY

compartment. one of the separate parts of a space that has been divided

cud. food that has been regurgitated

festival. a celebration that happens at the same time each year

regurgitate. to bring food that has already been swallowed back into the mouth

tradition. customs, practices, or beliefs passed from one generation to the next

unique. the only one of its kind

拉西的小木屋

绵羊

[美] 凯利·杜德娜 著
[美] 尼娜·乔拉 绘　王 妹 译

长春出版社
国家一级出版社
全国百佳图书出版单位

吉图字：07-2015-4474

图书在版编目（CIP）数据

我们身边的动物．拉西的小木屋·绵羊 /（美）杜德娜著；（美）乔拉绘；王妹译．— 长春：长春出版社，2016.1
（动物故事）
ISBN 978-7-5445-4173-2

Ⅰ．①我… Ⅱ．①杜… ②乔… ③王… Ⅲ．①绵羊—儿童读物 Ⅳ．①Q95-49

中国版本图书馆CIP数据核字（2015）第269761号

著　　者［美］凯利·杜德娜　　绘　　者［美］尼娜·乔拉
译　　者　王　妹　　责任编辑　张　岚　刘　洋　单紫薇

長春出版社 出版　　吉林省吉广国际广告股份有限公司印制
（长春市建设街1377号）（邮编130061　电话 88563443）　　长春出版社发行部发行
787㎜×1092㎜　20开　14.4印张　108千字
2016年1月第1版　2016年1月第1次印刷
定价：80.00元（全12册）

www.cccbs.net

真实&虚构故事

这套充满奇思妙想的图书专为学龄前后的小读者设计。它以极为巧妙的方式，在引领小读者探索自然科学世界的同时，循序渐进地加强了小读者的阅读能力，并以自如穿梭于真实世界与虚构故事间的形式，启发小读者的思维。

FACT 真实：左侧页面展示了真实照片，以增强小读者的科学认知和对信息文本的理解。

FICTION 虚构故事：右侧页面通过一个有趣的叙事故事，配合奇思妙想的插图，来启发小读者无穷的想象力。

FACT OR FICTION? 真实和虚构故事的页面可以分别进行阅读，通过提问、预测、推断和总结的方式提高理解能力。也可以并排连续阅读，能够帮助小读者发掘和研究不同的写作风格。

真实还是虚构故事？通过有趣的测验，加强小读者对于何为真实何为虚构的理解。

书后特附英文原文和词汇表，并配有外教的纯正发音，以极为便捷的方式让小读者感受地道的英文。

FACT

绵羊的幼崽被称为小羔羊。母羊每次能够产下1 到 3 只小羔羊。

小羊拉西觉得很无聊。“除了织毛衣，就没什么事情可做了。”她抱怨道。

“织毛衣就是我们应该做的事情呀。”拉西的双胞胎哥哥里欧安慰她说。

FACT

绵羊的体毛非常绵密。一只绵羊每年能产 4.5 千克左右的羊毛。

拉西穿上一件红色的格子衬衫。

“哇！穿上它，我就可以去当一名伐木工人啦！”拉西说。

“这是我听到的最愚蠢的话了！”里欧毫不留情地回应她。

FACT

羊喜欢群居。如果有人靠近它们，这会让它们很紧张。

“羊儿们就该和羊群待在一起。不然，在树林中落单的你会害怕的！”里欧对她说。

“不，我才不会害怕呢！”拉西大声说，“等雪丽去午睡的时候，我就偷偷溜出去！”

FACT

人们通常用狗来将羊群赶到一起。有时，人们也用骆驼来吓跑羊群周围的其他野生动物。

牧羊犬雪丽在牧场中四处巡视。

“一切看起来都很正常。好了，到午睡的时间啦！”她说着打了个哈欠。

雪丽在大树下舒展身体趴了下来，边打哈欠边数“一只羊、两只羊……”

FACT

羊的胃分四个部分。它们吃青草和其他杂草。比起静止的水，它们更喜欢喝流动的水。

拉西一直等到雪丽去午睡。

“我晚饭的时候就回来。”她对里欧说。

然后，她从围栏底下钻了出去，拉着她的四轮车上了山坡。

FACT

羊更喜欢爬上山坡去吃草。

拉西每天都用这种办法偷偷溜出来，她翻过小山，进入森林。

每天，她都会砍一棵小树；每天，她都往牧场运回一根原木。

FACT

羊能记住其他羊的长相。即便长达两年未曾见过面，它们还是能认出原来在同一羊群中的伙伴。

直到有一天，拉西让正在织毛衣的小羊们停下来，帮她收拾一下这些原木。

“你想建一个什么东西呢？”其他小羊好奇地问。

“你们会看到的！”拉西神秘地笑着回答。

FACT

天气太热或者太潮湿的时候，绵羊需要寻找遮蔽物纳凉。

拉西和羊群里其他的小羊们一起，在牧场的路边搭起了一个小木屋。

“好了！有了这个小木屋，不管刮风还是下雨，都可以卖我们织好的毛衣了！”拉西高兴地说道。

其他小羊也为此感到开心。

里欧跟着说道：“哇哦，拉西！我想下次你要是再想离开羊群去琢磨什么新点子，可一点都不成问题啦！”

FACT OR FICTION?

阅读下面的句子，判断它们出自真实的部分还是虚构故事的部分。

1. 绵羊的体毛非常绵密。

2. 羊能砍树。

3. 羊吃青草和其他杂草。

4. 羊在路边摊位上卖毛衣。

答案

1 真实 2 虚构 3 真实 4 虚构

FACT SPEED READ

A baby sheep is a lamb. A mother sheep is a ewe. 12
Ewes give birth to one to three lambs at a time. 23

A sheep's wool is called fleece. A sheep can grow up 34
to ten pounds of wool in a year. 42

Sheep like to be in a group, or flock. They get 53
nervous if they are by themselves. 59

Dogs are used to keep the flock of sheep together. 69
Sometimes llamas are employed to scare wild animals 77
away from the flock. 81

A sheep's stomach has four parts. Sheep eat grass 90
and weeds. They would rather drink running water 98
than still water. 101

Sheep prefer to walk uphill rather than downhill. 109

Sheep remember the faces of other sheep. They will 118
still know a flock mate that they haven't seen for up to 130
two years. 132

Sheep need shelter if the weather is too sunny or 142
too wet. 144

FICTION SPEED READ

Lacy Lamb is bored. "There's nothing to do except knit sweaters," she complains.

"Knitting is what we do," says her twin brother, Leo.

Lacy puts on a red plaid shirt. "Baa! I'm going to be a lumberjack," she says.

"That's the silliest thing I've ever heard!" Leo exclaims.

"Sheep are supposed to stay with the flock. You'll be scared in the woods by yourself," Leo says.

"No I won't," Lacy declares. "When Shirley takes her afternoon nap, I'm getting out of here!"

Shirley Sheepdog trots around the pasture. "Everything looks okay here. Nap time!" she woofs. She stretches out under a tree, yawns, and begins to count, "One sheep, two sheep ..."

Lacy waits until Shirley is snoring. “I’ll be back 120
by suppertime,” she says to Leo. She slips under the 130
fence and pulls her wagon up the hill. 138

Every day Lacy goes up the hill to the forest. 148
Every day she chops down another small tree. Every 157
day she wheels another log back to the pasture. 166

One day Lacy asks the other sheep to take a 176
break from knitting sweaters to help her arrange 184
the logs. “What will it be?” they wonder. 192

“You’ll see!” Lacy says with a smile. 199

Lacy and the flock stack the logs until they have 209
built a little roadside stand at the edge of the 219
pasture. 220

“Now we have a place to sell our sweaters, rain or 231
shine!” Lacy exclaims. The other sheep smile back 239
at her. 241

Leo says, “Wow, Lacy! I guess it’s okay to think 251
outside the flock after all!” 256

GLOSSARY

fleece. the coat of wool that covers some animals

flock. a group of animals or birds that have gathered or been herded together

llama. a smaller relative of the camel that is raised for its wool and is often used to carry loads across mountains in South America

pasture. land where animals feed on grass and other plants

shelter. protection from the weather

stand. a small, open building where things are sold

wool. the soft wavy or curly hair of animals such as sheep, llamas, and alpaca

弗兰妮的新鞋子

马

［美］ 南希·裘麦莉 著
［美］ C.A. 诺本斯 绘　　王 妹 译

長春出版社
国 家 一 级 出 版 社
全国百佳图书出版单位

吉图字：07-2015-4474

图书在版编目（CIP）数据

我们身边的动物．弗兰妮的新鞋子·马／（美）裘麦莉著；（美）诺本斯绘；王妹译．—长春：长春出版社，2016.1
（动物故事）
ISBN 978-7-5445-4173-2

Ⅰ．①我… Ⅱ．①裘… ②诺… ③王… Ⅲ．①马—儿童读物 Ⅳ．①Q95-49

中国版本图书馆CIP数据核字（2015）第269765号

著　者［美］南希·裘麦莉　绘　者［美］C.A. 诺本斯
译　者 王　妹　责任编辑 张　岚　刘　洋　单紫薇

長春出版社 出版　吉林省吉广国际广告股份有限公司印制
（长春市建设街1377号）（邮编130061　电话 88563443）　长春出版社发行部发行
787mm×1092mm　20开　14.4印张　108千字
2016年1月第1版　2016年1月第1次印刷
定价：80.00元
（全12册）
www.cccbs.net

真实&虚构故事

这套充满奇思妙想的图书专为学龄前后的小读者设计。它以极为巧妙的方式，在引领小读者探索自然科学世界的同时，循序渐进地加强了小读者的阅读能力，并以自如穿梭于真实世界与虚构故事间的形式，启发小读者的思维。

FACT 真实：左侧页面展示了真实照片，以增强小读者的科学认知和对信息文本的理解。

FICTION 虚构故事：右侧页面通过一个有趣的叙事故事，配合奇思妙想的插图，来启发小读者无穷的想象力。

FACT OR FICTION? 真实和虚构故事的页面可以分别进行阅读，通过提问、预测、推断和总结的方式提高理解能力。也可以并排连续阅读，能够帮助小读者发掘和研究不同的写作风格。

真实还是虚构故事？通过有趣的测验，加强小读者对于何为真实何为虚构的理解。

书后特附英文原文和词汇表，并配有外教的纯正发音，以极为便捷的方式让小读者感受地道的英文。

FACT

雌性马被称为母马。年幼的小马被叫作马驹。

明天将是小马驹弗兰妮第一天上学。

弗兰妮有些担忧地对妈妈玛尔说：“我不想去上学！我怕没人和我做朋友。”

FACT

马脖子上的毛发被称作鬃毛。有些马的鬃毛很长，有些马的鬃毛很短。

弗兰妮在池塘的水中照了照自己。

“如果我把头发编起来，再染成紫色，那大家都会觉得我很酷！就会跟我交朋友了。”

FACT

马能够同时看到身体两边的事物。但是，马看不到它们正前方的东西。

“如果我戴上闪闪发亮的紫色项链，那多时髦呀，每个人都会喜欢我。”弗兰妮说。

FACT

马用尾巴来驱赶飞蝇。它们也通过尾巴来回摆动与其他的马交流。

妈妈摆着尾巴对她说："不论你长相和穿戴如何，都要过得开心！大家都会喜欢真正的你，想和你做朋友。做你自己就好了，我的宝贝。"弗兰妮想了一会儿说："我想你是对的！"

FACT

马吃青草、燕麦和干草。它们也会吃胡萝卜和苹果。

“我饿了！午餐可以吃汉堡和炸薯条吗？”

“可以呀！”妈妈说，“那我们晚餐就喝蔬菜汤好了！”

说着，她们向马房跑去。

FACT

不同种类的马身上有不同的图案。其中有一种马腿上的图案看起来像袜子一样。

“你的鞋子看起来已经小了呀！”妈妈说，“午餐后，我们一起去商店吧！或许我们能够找到合适的鞋子，来配你最喜欢的袜子。”

FACT

马蹄和指甲类似，必须经常进行清洁和修剪。很多马的马蹄会被套上金属马掌，起到保护作用。

她们去了五家商店，最后才找到完美的鞋子。

穿着蓝橙相间的高跟马鞍鞋，弗兰妮一边高兴地说“这正是我想要的！”，一边昂首阔步地展示她的新装。

“新鞋子跟我的橙色波尔卡圆点中筒袜很相配，真是棒极了！”

FACT

马不喜欢被人轻拍，它们更喜欢被人轻抚或用手掌来回摩挲。

那天晚上，妈妈给小弗兰妮盖好被子，抚摸着她的背，陪她入睡。

“一想到明天要去上学，我就很兴奋。我希望午餐能吃到汉堡和炸薯条。”

弗兰妮说着闭上了眼睛。

“还有，谢谢妈妈，谢谢你为我买的新鞋子！”

FACT OR FICTION?

阅读下面的句子，判断它们出自真实的部分还是虚构故事的部分。

1. 马用尾巴驱赶飞蝇。

2. 马可以同时看到身体两侧的事物。

3. 马会去商店买鞋子。

4. 马吃汉堡和炸薯条。

答案

1 真实 2 真实 3 虚构 4 虚构

FACT SPEED READ

A female horse is called a mare. A baby horse is 11
called a foal. 14

The hair on a horse's neck is called a mane. Some 25
manes are long. Some manes are short. 32

Horses can see in two directions at once. But horses 42
cannot see right in front of them. 49

Horses use their tails to swat flies. Horses also swish 59
their tails to communicate with other horses. 66

Horses eat grass, oats, and hay. Horses eat carrots 75
and apples too. 78

Horses have many kinds of markings. One marking 86
looks like socks. 89

Horses' hooves grow like toenails and must be 97
cleaned and trimmed. Many horses wear metal shoes 105
to protect their hooves. 109

Horses do not like people to pat them. Horses like 119
to be rubbed and stroked instead. 125

FICTION SPEED READ

Tomorrow is the first day of school. Frannie 8
Foal says to Mama Mare, "I don't want to go to 19
school! I'm afraid that no one will be my friend." 29

Frannie looks at herself in the pond and says, 38
"If I braid my hair and dye it purple, everyone 48
will think I'm cool and will want to be my 58
friend." 59

"If I wear big purple-glitter glasses, then I'll look 69
hip and everyone will like me," Frannie says. 77

Mama swishes her tail and says, "Have fun 85
with what you look like and what you wear! But 95
others will like you and want to be your friend 105
for who you are. Just be yourself, Frannie." 113

Frannie thinks for a moment. "I guess you're 121
right," she says. 124

"I'm hungry! May I have a burger and fries for 134
lunch?" Frannie asks. 137

"Okay," Mama says, "but it's veggie soup for 145
dinner!" They gallop back to the barn. 152

"Your shoes do look small," Mama says. "After 160
lunch, we'll go shopping. Maybe we can find some 169
shoes that will be fun to wear with your favorite 179
socks." 180

They go to five stores before Frannie finds the 189
perfect shoes. 191

Frannie exclaims, "These are really me!" as she 199
prances around in blue-and-orange high-heeled 207
saddle shoes. "They look great with my orange 215
polka-dot knee-highs!" 219

That night, Mama tucks Frannie into bed and 227
rubs her back. "I'm excited to go to school 236
tomorrow. I hope we have burgers and fries for 245
lunch," Frannie says as she closes her eyes. "And 254
thank you, Mama, for the new shoes!" 261

GLOSSARY

hoof. the hard covering that protects the foot of an animal such as a horse, cow, or deer

knee-high. a sock or stocking that covers the foot and leg up to the knee

marking. the usual pattern of color on an animal

pond. a body of water smaller than a lake

prance. to walk in a lively, springy way

saddle shoe. a laced, leather shoe that has a black or colored band of leather across the instep